Contents

Contents

Applied Principles of Horticultural Science

L. V. Brown
BSc (Hons), PgD (LWM), AMIAgrE, MISoilSci, FRGS, Cert Ed

Butterworth-Heinemann
Linacre House, Jordan Hill, Oxford OX2 8DP
A division of Reed Educational and Professional Publishing Ltd

 A member of the Reed Elsevier plc group

OXFORD   BOSTON   JOHANNESBURG
MELBOURNE   NEW DELHI   SINGAPORE

First published 1996

British Library Cataloguing in Publication Data
Brown, L. V.
 Applied principles of horticultural science
 1 Horticulture   2 Soil Science
 I Title
 635

ISBN 0 7506 2954 1

Composition by Genesis Typesetting, Laser Quay, Rochester, Kent
Printed and bound in Great Britain

Part Two Soil science

Contents

Preface

Aims

This is a completely new handbook, designed for people who wish to explore and investigate the natural resources of the land-based technologies and, through this, apply their science knowledge in practical professional situations. It is intended to enable the learner competently to carry out routine scientific applications of horticultural principles.

The book is best regarded as a method to import practical understanding to the learner in a 'hands-on' environment, very much on the basis that 'an ounce of practical is worth a tonne of theory'. This is approached through relating and presenting the science principles of plants, soils and organisms, as an interactive applied process. There is therefore, in one textbook, coverage of most of the primary science used in the land-based technologies.

Approach to the subject

The handbook assumes no scientific knowledge above compulsory school education. This book differs from others in that special emphasis is laid on evidence gained from practical experimental studies. This is integrated with knowledge of the theoretical concepts gained from the supplementary text sections. Additional support which may be gained from: Adams, Bamford, and Early (1995).

Through observing and applying the science of horticulture the learner may increase understanding of how plants grow to become mature, healthy specimens established in either soils or containers. Practical exercises enhance study points and greater topic exploration emphasizes their relevance to growing plants, thus deepening core knowledge.

The book presents exercises succinctly, in the context of their relationship within the horticultural environment. The learner should make his or her own notes and drawings since this practice can aid absorption and understanding in a relaxed informal manner. The presence of questions supports independent learning and private study; they enhance the link to related studies and check that understanding has taken place.

Overall structure

The book is divided into three parts: Plant science; Soil science; Pest and disease.

Each part contains a number of chapters relating to a major principle of applied horticulture (e.g. Propagation; Soil water; Insects and mites).

Chapters are further partitioned into exercises which contain applied skills necessary to manipulate and manage the horticultural resources (e.g. photosynthesis, flower structure, raising soil pH, fertilizers, etc.). Skilful management of such resources is increasingly important in today's environment, where the difference between commercial success and failure rests largely on the speed and accuracy with which the horticulturalist completes the job. Increasingly, responsibility is being delegated to lower levels of staff and the need for skilled operatives is more apparent.

Each of these exercises is composed of practical tasks containing a very specific skill or competence, with which the learner engages, together with the supporting exercises, as a method to advancing their skills in the area under study.

Only the barest theory is included reflecting the handbooks emphasis on 'doing' skills. Wherever practicable, topics are approached from the horticultural angle with the underpinning applied principles slotted in.

Each exercise clearly sets out a logical learning sequence containing:

Background	brief introduction to the theme and context
Aim	purpose and justification for doing the activity
Apparatus	list of resources required to complete the activity
Method	a logical sequence of data obtaining tasks
Results	tables to complete, thus ensuring all learning outcomes are addressed
Conclusions	Problem-solving transfer of information gained to practical real world environment, or new situations.

A complete list of all exercises can be found in the Contents pages.

Throughout the text definitions of important terms are tinted.

Purpose

The purpose of this book is to provide a repertoire of practical investigations and student-centred learning study questions for the benefit of both instructors/teachers and learners. Greater emphasis is now placed on learning skills in the workplace. Modern apprenticeships and National Vocational Qualifications (NVQs) are examples of this and are well suited to the approach of this book.

Manipulation of plant yield comes from a correct mastery of the interacting environmental variables (soils, organisms, plants). Through increased understanding and management of both plants and the resource of the land, the land-based technologist can substantially increase the ease of plant cultivation and profit. The investigative exercises offered in this book develop learner awareness and competence in the application of these principles to achieving this aim.

In recent years there has been rapid progress made in both teaching methods and student learning styles. Learners are better able to negotiate which lectures they attend, at what times and how long their course will be. Modularization of course structures has also enabled the learner to take only one or two modules per year and to advance at their own pace.

This handbook has been designed against this background of a more 'student centred' self-managed learning environment. For example, it may be that theoretical components are covered by reference to specific books or lectures and the learner follows up by independent experimentation in the laboratory, workplace or other learning environment.

The author and publishers do not accept liability for any error or omission in the content, or for any loss, damage or other accident arising from the use of information stated herein. Inclusion of product or manufactures information does not constitute an endorsement and in all cases the manufacturer's instructions and any relevant legislation pertaining to the use of such materials should be followed.

I lay no claim to originality. Some of the subjects have been attempted in various different ways before. A few questions have been adapted from other books within the biological sciences. Suggestions for improvements and additional exercises would be gratefully received via the publishers in the first instance.

L. V. Brown

Acknowledgements

I am indebted to the following people for their support and encouragement in the preparation of this book. Mike Early, for initially encouraging and motivating me to produce this publication. Dr Nigel Scopes, Bunting Biological Control Ltd, for comments on Biological Control exercises. Carol Oldnow, for some technical data. Matthew Deans and the reviewers from Butterworth-Heinemann publishers. Slides of biological control organisms were kindly supplied by the following: Sue Jupe, Public Relations and Marketing Services, Defenders Ltd; Dr Mike Copland, Wye College; and Peter Squires, Senior Entomologist Koppert Biological Systems. Finally, my parents, Sid and Audrey Brown, for providing a suitable environment to enable me to work undisturbed on the initial manuscript.

Part One   Plant science

1 Plant kingdom classification and nomenclature

Background

Living organisms may be grouped simply into two kingdoms; plants or animals. Plants are the major group of living organisms that have their cells surrounded by cell walls built of fibres. They are also the group of primary importance in horticulture.

All the higher plants, and many others, have evolved from green algae and contain pigments such as chlorophyll that can trap solar radiation energy, to make energy-rich carbohydrates by photosynthesis. The non-green plants such as bacteria and fungi do not normally contain chlorophyll and have to obtain their energy-rich compounds from other organisms. This type of organism is called a parasite.

The plant kingdom is extremely large – estimated at three hundred thousand different species – and if fungi are included this rises to an estimate of four hundred and fifty thousand. This includes two hundred and fifty thousand species of flowering plants alone.

It is therefore helpful to classify plants into similar groups and to name plants with a name that is understood all over the world. It is generally useful to classify plants into **lower** and **higher** plants. Lower plants reproduce by spores. Higher plants reproduce by seed. Using this knowledge, six simplified divisions of the plant kingdom offer sufficient detail to correlate plants with their significance in horticulture (see Table 1.1).

The wide variety of plant types found represent different evolutionary pressures in their natural environment. Plants can be grouped fairly easily by characteristics or modes of life. Such a classification system enables a horticulturalist to identify any plant from any country. It may also enable predictions about the behaviour of unknown plants from knowledge of similar plants

Division	Significance in horticulture
Lower plants (reproduce by spores)	
Chlorophyta (algae)	very simple plants indicator of damp conditions indicator of nitrogen and causes pollution in ponds and streams sand filters are used to remove algae from irrigation waters
Mycophyta (fungi)	very simple plants containing groups of cells in threads recycle organic matter and nutrients source of disease and parasites
Lichenes (lichens)	very simple plants a symbiotic relationship between algae and fungi they are indicators of clean air/pollution often used for decorative effect (rural planning authorities have been known to insist that rebuilt church walls are smeared with a mixture of yoghurt and dung to stimulate lichen growth!)
Bryophyta (mosses and liverworts)	plants which have stems and leaves source of peat/compost for container-grown plants decorative use smother out grass indicates dampness sometimes called the amphibians of the plant world
Pteridophyta (ferns and horsetails)	plants which have stems, roots and leaves ferns make pots and garden specimens and like a shady position horsetails are problem weeds not responding to weedkillers
Higher plants (reproduce by seed)	
Spermatophyta (seed bearing)	The most important group in horticulture. Plants with stems, roots, leaves and a transport system. They may be either gymnosperms (cone bearing, the seed is born 'naked', e.g. conifers) or angiosperms (flowering plants with seed within a fruit either monocotyledonous or dicotyledonous)

Table 1.1
Plant kingdom divisions

from the same group (e.g. Rosaceae family are all susceptible to fireblight).

Examples of the use of the classification system world wide are given in Table 1.2. Note particularly the word endings (e.g. 'aceae' for family), which are standardized and also that genus and species appear in *italics* (or alternatively, may be <u>underlined</u>).

	Example usage	
	Scientific terms	*Common terms*
Kingdom	Plant**ae**	Plants
Division	Spermato**phyta**	Seed bearing plants
Class	Angiosperm**ae**	Flowering plants
(sub class)	Dicotyledon**ae**	Dicotyledon
Order	Ran**ales**	Ran**ales**
Family	Ranuncul**aceae**	Ranuncul**aceae**
Genus	*Ranunculus*	*Ranunculus*
Species	*repens*	*repens*
Common name	Creeping buttercup	Creeping buttercup

Table 1.2
Classification system usage

A simple classification map is given in Figure 1.1.

Figure 1.1 demonstrates how a classification map is constructed and provides an overview of the main points in classification and nomenclature. These areas are subject to opinion and consensus with no overall authority.

Division: a group of related classes. Names should usually end **phyta**.

Class: related orders form a class, mostly ending **ae**.

Order: closely related families make an order, usually ending **ales**.

Family: a group of genera which contain plants with similar flowers, but a wide variety of growth forms and characteristics (trees, shrubs, perennials, etc.). The family Rosaceae contains a wide range of genera (e.g. Rosa, Malus, Prunus, Cotoneaster, Crataegus, Pyracantha, Rubus, Sorbus and Potentilla). However, nearly all the plants in the Rosaceae family suffer from fireblight. Family names should end in **aceae**.

The Binomial System, designed by Carlos Linnaeus, an eighteenth-century Swedish botanist, prevents confusion and must be obeyed when naming plants following the rules of the international code of botanical nomenclature, and the international code for cultivated plants. Latin is the universally

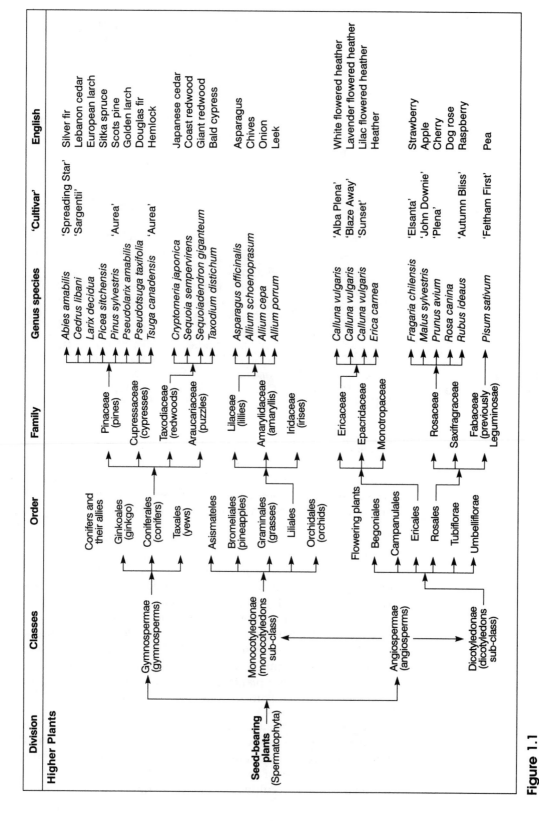

Figure 1.1

Simplified classification of the plant kingdom showing some example pathways

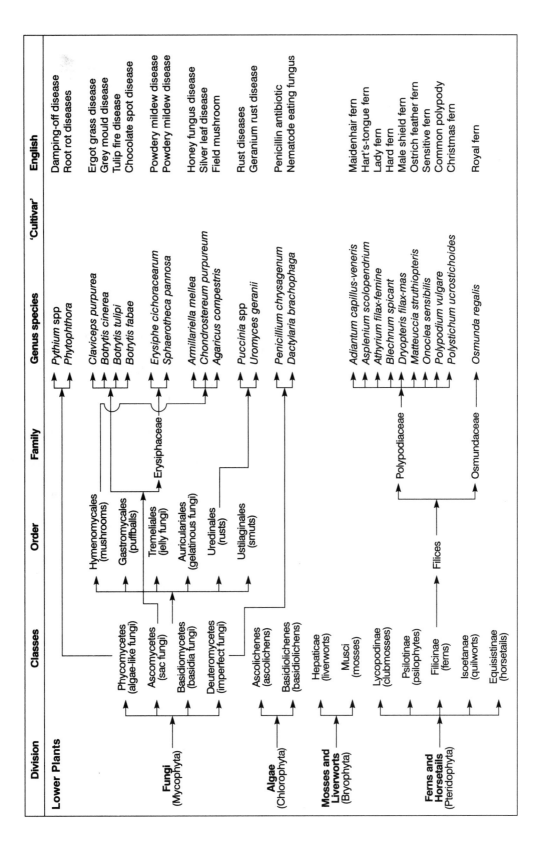

accepted scientific language for the specification and naming of these groups. Although it is a dead language not spoken by any nation it should be understood worldwide. The names must have a genus and species, which must be in Latin (e.g. *Chrysanthemum morifolium* – Chrysanthemum).

Genus and species names should be in Latin and either *italic* or underlined. A genus: (genera) is a group of closely related species. Always start with a capital letter. The species is the basic unit of plants that can interbreed but not with members of another species. Always start with a lower case letter.

Cultivars are **culti**vated **var**ieties, bred for a particular characteristic. They begin with a capital letter and should be in a modern language, not Latin, with single inverted commas ' ' or the term **cv**.

It is important that these rules and standards are followed when naming or specifying plants as they enable speedy ordering of goods from around the world in a scientific language that all professional horticulturalists use. Much of the nursery stock and cut flower products are imported into Great Britain from Aalsmeer Flower Auctions in Holland, where all goods are bought by specifying cultivar and using Latin names.

To horticulturalists, classification above 'Family' seems unimportant. Family names themselves can be of great value. For example, when choosing herbicides and pesticides; or identifying plants by flower shape to aid crop rotations and reduce plant susceptibility to pest attack.

The higher plants are divided into conifers and their relatives (gymnosperms), and flowering plants (angiosperms). Both groups reproduce by producing a seed. However in the gymnosperms the seeds are carried 'naked' rather than enclosed in a fruit (as in angiosperms). The seeds are often borne into cones, but may sometimes be surrounded by fleshy outgrowths from the stem. For example *Taxus baccata* – the yew. Included in this group are relatives of the conifers, such as *Ginkgo biloba*, found as a 'living fossil'.

Exercise 1.1 *The plant kingdom*

Background

The plant kingdom may be conveniently divided up into groupings of plants as they affect horticulture. The characteristics and patterns of these divisions has been described in this chapter.

Aim

To investigate the division of the plant kingdom and its effect on horticulture.

Apparatus

Introductory text.

List of terms:

algae	higher plants
angiosperm	lichen
dicotyledons	lower plants
ferns and horsetails	monocotyledons
fungi	mosses and liverworts
gymnosperm	

Method

Match the correct terminology to the sentence describing the feature.

Results

Write the correct term next to each statement.

1 Plants that reproduce by seed.	
2 These plants reproduce by spores.	
3 Seed bearing plants with naked seed.	
4 These organisms have their cells arranged in threads and do not contain chlorophyll.	
5 Plants with stems, roots and leaves but not transport vessels, often weeds.	
6 Plants with seeds having one seed leaf.	
7 A division mostly used in decorative horticulture.	
8 A group of plants mostly used as a growing medium.	
9 The presence of these plants in irrigation water may indicate pollution.	
10 Seed bearing plants with seed contained in a fruit.	
11 Plants with seeds having two seed leaves.	

Conclusions

1 State the common name for the following divisions of the plant kingdom.
>Bryophyta
>Chlorophyta
>Lichenes
>Mycophyta
>Pteridophyta
>Spermatophyta

2 Describe the plant characteristics botanists use to classify lower and higher plants.

3 Why are ferns slightly more adapted to life on land than mosses and liverworts?

4 State two situations where lower plants are used in horticulture.

Exercise 1.2 *Principles of classification*

Background

The divisions of the plant kingdom are divided into plants with similar characteristics. They are classified into units of increasing specialism from 'Division' right down to 'Cultivar'. There are rules for the way to write down plant names, including word endings (such as the binomial system) and this may help to aid recognition of plant characteristics with which you may be unfamiliar.

Aim

To investigate some horticultural plants as a method to increase understanding of classification.

Apparatus

Introductory text.
List of terms found with each statement.

Method

Read and follow the instructions for each of the statements.

Results

Enter your result in the table provided.

---

1 Number the following terms into logical structured order:

| Kingdom | Order | Class | Species | Sub-class | Family | Genus | Division | Cultivar |

---

2 In the following four plants state which parts of their names are:

	Apple	Bean	Hollyhock	Annual meadow grass
(a) Kingdom				
(b) Division				
(c) Class				
(d) Order				
(e) Family				
(f) Genus				
(g) Species				
(h) Cultivar				

Apple: Plantae Spermatophyta Dicotyledonae Rosales Rosaceae *Malus domestica* 'Golden Delicious'
Bean: Plantae Spermatophyta Dicotyledonae Rosales Fabaceae *Phaseolus vulgaris* 'Stringless Green Pod'
Hollyhock: Plantae Spermatophyta Dicotyledonae Malvales Malvaceae Altheae rosea 'Newport Pink'
Annual meadow grass: Plantae Spermatophyta Monocotyledonae Graminales Poaceae Poa annua

3 With reference to Figure 1.1 construct a classification map for the following plants:

    (a) Dog rose
    (b) Pea

4 Name endings are often useful to help identify the classification of unfamiliar plants. State which of the following terms are : (a) Family and (b) Order.

Aceraceae	Cupressaceae	Lilaceae	Pinaceae	Taxales
Asteraceae	Ericaceae	Magnoliales	Poaceae	Taxodiaceae
Begoniaceae	Ericales	Orchidales	Rosaceae	Urticaceae
Cannabaceae	Graminales	Palmaceae	Solanaceae	Urticals
Coniferales				

5 There are nearly 100 separate Family names. Observe the following list of plant families.

Asteraceae	(previously Compositae)	– the daisy family
Brassicaceae	(previously Cruciferae)	– wallflower family, including many vegetables
Hyperiaceae	(previously Guttiferae)	– St John's Wort family
Lamiaceae	(previously Labiatae)	– mint family, often aromatic
Apiaceae	(previously Umbelliferae)	– parsley family, often aromatic
Fabaceae	(previously Leguminosae)	– peas and beans family
Poaceae	(previously Graminae)	– grass family

Using your knowledge of classification and nomenclature:

(a) Explain the rules that made it necessary to rename these seven families.
(b) Explain why some modern books may be in error by continuing to use old family names.

Conclusions

1 Describe the importance of Family names in horticulture.
2 State, giving one example in each case, the correct word endings for:

(a) Division
(b) Class
(c) Order
(d) Family

3 State two ways in which plant classification helps horticulturists.

Exercise 1.3 *Nomenclature for plant ordering*

Background

The binomial system sets out the rules by which plants should be named. The effect of this in horticulture is upon plant ordering in what is now a global market, whether it be importing cut flowers from Holland, semi-mature trees from Italy, chrysanthemum cuttings from the Canary Islands, Christmas poinsettias from America, Cape fruit from South Africa, vegetables from Kenya or hedging plants from the local garden centre. The internationally recognized language is Latin. It is helpful, but not essential, to specify the Family name. Genus and species should always be stated although sometimes ordering by cultivar is sufficient (but is technically incorrect). This is often done when we know precisely the effect a cultivar would give, such as when ordering cut flowers:

e.g. *Alstroemeriaceae Alstroemeria aurantiaca*

'Capri' – pale pink flowers
'Stayelar' – striped pattern flowers
'Zebra', 'Orchid', 'Canaria' – orchid type flowers
'Rosario', 'Jacqueline' – butterfly type flowers

Aim

To gain confidence in the use of plant nomenclature to order plants accurately.

Apparatus

Introductory text.
List of terms appearing in each section, Figure 1.1.

Method

Follow the instructions given in each of the statements below.

Results

Enter your observations in the table provided.

1 For each of the following plant names state which refers to:

	Scarlet pimpernel	Transvaal daisy	Yew	Snap dragon	Gilly flower
(a) the species					
(b) the family					
(c) the genus					
(d) the cultivar					

Primulacaeae	*Anagallis arvensis* 'Golden Yellow'	– Scarlet Pimpernel
Asteraceae	Gerbera jamesonii	– Transvaal Daisy
Ericaceae	*Calluna vulgaris* 'Alba Plena'	– summer white flowering heather
Taxodiaceae	*Taxus baccata* 'fastigiata'	– Yew
Scrophulariaceae	Antirrhinum majus 'His Excellency'	– Snap dragon
Mathiola incana	'Francesca'	– red flowered Gilly flower

2 Which **one** of the following orders is correctly written for Antirrhinums (snap dragon)

    (a) Scrophulariaceae *Antirrhinum majus* "His Excellency"   (intermediate flowers)
    (b) *Antirrhinum Majus* "His Excellency"   (intermediate flowers)
    (c) Scrophulariaceae Antirrhinum majus 'Coronette'   (tall flowers)
    (d) Scrophulariaceae antirrhinum majus 'little darling'   (dwarf flowers)
    (e) *Antirrhinum* majus 'Coronette'   (tall flowers)
    (f) *Antirrhinum majus* 'his Excellency'   (intermediate flowers)

3 State what is wrong, if anything, with each of the following seed orders.

    (a) Poaceae *Alopecurus pratensis* 'Aureus' (golden foxtail)   (an ornamental grass)
    (b) Alopecurus pratensis 'Aureus'
    (c) Alopecurus pratensis "Aureus"
    (d) *Alopecurus pratensis* golden foxtail
    (e) Poa triviales   (a lawn grass)
    (f) Rutaceae *Citrus Limonia*   (lemon)
    (g) *Citrus Limonia*
    (h) *Citrus limonia*
    (i) Solanaceae Lycopersion esculentum   (tomato)
    (j) Solanaceae lycopersion esculentum

4 The following order was received by a local garden centre to supply immediate summer colour to a small garden project. List all **six** technical errors.

(a) Transvaal Daisy (Gerberas): Asteraceae *Gerbera Jamesonii*
(b) Gilly Flowers:

Mathiola incana	``Francesca''	(red flowers)
Mathiola incana	`arabella'	(lavender flowers)
	`Debora'	(purple flowers)

(c) Summer flowering heathers:

Ericaceae <u>Calluna vulgaris</u>	`Alba Plena'	(white flowers)
Calluna vulgaris	`Alba Rigida'	(white flowers)
Calluna Vulgaris	` Allegro'	(purplish-red flowers)

Conclusions

1 State the meaning of the term 'species'.
2 State the correct word endings for 'Family' names.
3 In which language should the genus and species names be written?
4 What variations should be made between writing genus and species?
5 Explain the correct way to order by cultivar.

Exercise 1.4 *Flowering plants (angiosperms)*

Background

The flowering plants (angiosperms) are divided into the two sub-classes called monocotyledon and dicotyledon and may be visually identified quite easily using the following guidance.

Flowering plants

The focus of most aspects of horticulture revolves around the growth and development of flowering plants (angiosperms). These always have developing seeds enclosed in a fruit and flowers, highly adapted for insect or wind pollination. They are divided into two sub-classes: the **mono-cotyledons** and the **dicotyledons** (some times abbreviated to monocot and dicot respectively). They have many differences which can effect our management of these plants (e.g. it is difficult to take cuttings from monocotyledons).

Dicotyledons

('Di' means two, 'cotyledon' means seed leaf)

These are plants bearing two cotyledons or seed leaves in the seed (e.g. broad leaf trees, roses, daisies, cabbages, pelargoniums, cacti, oak, tomato).

All seeds contain a juvenile plant embryo and a food reserve normally as a cotyledon. The cotyledon supplies energy until the germinating plant is able to photosynthesize.

The seed leaves or cotyledons may emerge above the ground with the shoot. The seed leaves turn green and can be confused with the true leaves still to expand from the shoot. Sometimes they remain below the ground, supplying their food reserve to the developing seedling.

Monocotyledons

('Mono' means one, 'cotyledon' means seed leaf)

This refers to a plant bearing only one cotyledon or seed leaf when they germinate (e.g. grass, lilies, orchids, palms, bromeliads, daffodils, crocus, palm tree, spider plant, pineapple, wheat, barley, oats, iris, onions, gladiola, tulip, narcissus).

They have parallel veins rather than the net-like veination of the dicotyledons. The flowers of the monocotyledons are in threes or in multiples of three, as opposed to dicotyledons with their flower parts in fours or fives. It is very rare for monocotyledons to form shrubs or trees because they don't form a cambium ring (growth tissue) to allow stem thickening.

The essential differences between monocotyledons and dicotyledons are summarized in Table 1.3:

Table 1.3

Distinguishing features between monocotyledonous and dicotyledonous plants

Plant tissue	Dicotyledon	Monocotyledon
Embryo seed (Exercise 2.1)	2 seed leaves (cotyledons)	1 seed leaf (cotyledon)
Roots (Exercise 4.3)	Primary root often persists and becomes strong tap root with smaller secondary roots Cambium present Xylem in a star shape	Primary roots of short duration, replaced by adventitious root system No cambium Xylem in a ring
Growth form	Herbaceous or woody	Mostly herbs, few woody
Pollen (Exercise 9.4)	Has 3 furrows or pores	Has 1 furrow or pore
Stem vascular bundles (Exercises 8.1 and 8.2)	Arranged in a ring round the stem Cambium present Secondary growth in stem Stem differentiated into cortex and stele	Scattered around stem No cambium No separation into cortex and stele
Leaves (Exercise 5.1)	Net veined Often broad No sheathing at base Petioles (stalk) present	Parallel veins Oblong in shape Sheathing at base No petiole (stalk)
Flowers (Exercises 9.1, 9.2 and 9.3)	Parts usually in multiples of 4 or 5	Parts usually in multiples of 3

Aim

To identify monocotyledon and dicotyledon sub-classes based upon visual observation.

Apparatus

Previous text section, Table 1.3

cabbage	lily
chrysanthemum	pelargonium
crocus	rose
daffodil	spider plant
geranium	

Method

Describe the specimen plants' characteristics under the headings given below.

Results

Enter your observations in the table provided.

Plant tissue being described	Description of plant tissue in a range of plants						
	cabbage	crocus	geranium	lily	pelargonium	rose	spider plant
Leaves (net veins or parallel veins?)							
Flowers (petals in multiples of 4/5 or 3s?)							
Roots (Tap or fibrous?)							
Growth form (herbaceous or woody?)							
Deduction: Monocotyledon or Dicotyledon?							

Observe Plant A and Plant B State, giving reasons, whether each is a member of the sub-class monocotyledon or dicotyledon		
Specimen	Sub-class	Reasons
Plant A, Daffodil		
Plant B, Chrysanthemum		

Conclusions

1 Explain the meaning of the terms 'monocotyledon' and 'dicotyledon'.
2 List four differences between monocotyledons and dicotyledons and name two examples of each.
3 State one significance in horticulture of the practical difference between the two sub-classes.

2 Seed viability and vigour

Background

Flowering plants begin their life cycle as seeds. The seed contains a miniature plant called an embryo, from which a new adult plant will grow. The young plant that develops from a germinating seed is called a seedling.

Seed quality (viability and vigour)

Most edible horticultural crops and bedding plants are grown from seed (see Chapter 3). In selecting seed to grow, quality becomes an important aspect. Our aim is to produce synchronized germination, uniform crop characteristics (e.g. height) and healthy, robust growth.

There are several properties that are used to help assess the quality of seed, such as:

- purity   percentage of undamaged seeds
- germination   percentage of viable seed
- health   presence of potential pathogens.

For edible horticultural crops and forest trees, the **International Seed Testing Association** governs the test procedure for these three characteristics, under the authority of the **Seed Acts**. A minimum germination percentage must be achieved. Flower seeds are not covered by the act. Germination percentage (**viability**), will vary under different environmental conditions. For example, more seed is likely to germinate in sheltered lowlands than in the cold high altitude of mountains and there is considerable variability in performance. This variation in performance is called **vigour** and this property is now regulated by the **Vigour Test Committee** of the International Seed Testing Association. In North America the **Vigour Committee of the Association of Official Seed Analysts of North America** performs a similar role. Our interest in vigour lies in knowledge of the proportion of seedlings established in the field and the rate of uniformity in their emergence.

In selecting seed for horticultural applications the buyer must be sure that the seed will germinate successfully (viability) and will succeed under a variety of hostile environments, e.g. cold (vigour).

Viable

Seeds which retain their ability to germinate. The older the seed, the less likely it is to be alive, therefore fewer seeds from old stock will germinate than seeds from new stock.

Vigour

Estimates the ability of the seed to germinate and develop into strong seedlings under a wide range of simulated field conditions, e.g. cold and hot.

The exercises in this section will explore these areas and include:

- seed structure and tissue function
- percentage germination and effect of seed age (viability test)
- embryo chemical reaction test for viability
- effect of different environments on germination (vigour test).

Exercise 2.1 *Seed structure and dissection*

Background

The tissue within the seed is adapted to ensure that the embryo plant will emerge at a time to ensure maximum favourability for growth and development. For example, the tough testa seed coat prevents germination until it is broken down by the harsh winter climate. Germination is therefore prevented until the environmental conditions of spring are suitable.

Flowering plants are divided into the two sub-classes; monocotyledon and dicotyledon. This refers to the number of seed leaves (called cotyledons) that the seedling establishes. Dicotyledonous plants have two cotyledons. During germination they draw on food that is stored in these seed leaves. Monocotyledonous plants have only one cotyledon and their food is stored in tissue called an endosperm. Figure 2.1 typifies the internal tissues of seeds of monocotyledon and dicotyledon plants.

19

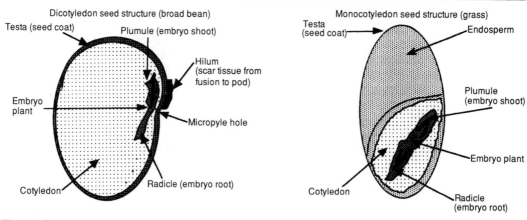

Figure 2.1
Internal seed tissue

Aim

To practise dissection and seed tissue identification.

Apparatus

bean seeds   scalpel
grass or wheat seeds   seed structure illustration
hand lens   white dissection tile

Method

1 Soak 20 seeds of each plant in water overnight to soften the tissues.
2 Bisect the bean seed and dissect the grass seed as shown in Figure 2.1.
3 Draw a labelled diagram of all the tissues found.

Results

Draw a labelled diagram of the observed seed stuctures.

Conclusions

1 What can you conclude about:

(a) seed dissection
(b) observation of tissue within the seed?

2 Name the tissue coating containing the seed.
3 State the name for the scar tissue formed where the seed was previously attached to the pod.
4 State the technical name for seed leaves.
5 What is the embryo shoot called?
6 What is the embryo root called?
7 Distinguish between seed viability and vigour.

Exercise 2.2 *Percentage germination*

Background

After the seed swells, having absorbed water through the micropyle through a process called osmosis (see Chapter 7), germination occurs. The radicle (or primary root) emerges from the seed, followed by the plumule and cotyledons. The cotyledons (seed leaves) supply food to the growing plant embryo for the production of new cells. The new plant will manufacture its own food by the process of photosynthesis as soon as it has grown tall enough to put out leaves (true leaves). Until then the cotyledons normally emerge above ground with the growing plumule, supplying energy. As their food reserves are used up the cotyledons shrivel and fall off.

Aim

To gain competence in calculating percentage germination.

Apparatus

    calculator
    compost tray
    oil seed rape seedlings

Method

1 Sow the compost tray with 40 seeds of oil seed rape.
2 Allow 7 days for germination to take place.
3 Count the number of seeds that have germinated.
4 Calculate the percentage germination rate using the following formulae:

$$\text{Percentage germination rate} = \frac{\text{number of successful seeds germinated}}{\text{total number of seeds sown}} \times 100$$

Results

Enter your results in the table provided.

Number of seeds sown	Number of seeds germinated	Percentage germination

Conclusions

1 Through which tissue does water enter the seed prior to germination?
2 What can you conclude about the viability of the oil seed rape seed?
3 How many cotyledons are present in the growing seedlings?
4 Explain whether oil seed rape is a monocotyledonous or dicotyledonous species.
5 State the functions of the cotyledons.

Exercise 2.3 *Effect of seed age on germination*

Aim

To investigate the effect of seed age on germination success.

Apparatus

calculator   old and new bean seeds
filter paper   old and new lettuce seeds
4 petri dishes

Method

1 Place a disc of filter paper in a petri dish and damp it down with water.
2 Sow petri dishes A and B with 10 old and new lettuce seeds respectively.
3 Sow petri dishes C and D with 10 old and new bean seeds respectively.
4 Monitor and keep the filter paper moist over the next 7 days.
5 After 7 days count the total number of seeds sown and calculate the percentage germination rate.

Results

Express your results in the table provided.

Petri dish	Number germinated	Total seeds sown	% germinated
A			
B			
C			
D			

Conclusions

1 Is this test for viability or vigour?
2 What can you conclude about the effect of seed age on germination?
3 How might your results influence your choice of seed storage conditions?

Exercise 2.4 *Tetrazolium test*

Background

Conducting germination tests on seeds can be time consuming. As an alternative method to sowing seeds and recording resulting growth, a chemical test may be used.

The embryos of seeds contain enzymes responsible for aiding germination. These enzymes react with a colourless 1 per cent triphenyl tetrazolium chloride solution to produce a brilliant red stain. This is referred to as the 'tetrazolium test'. Seeds that will germinate normally are stained over the whole embryo from shoot to root. The more enzymes that are present, the greater and more widespread is the red stain, indicating germination potential.

The method is fast, and results can be obtained normally within twenty-four hours. The test is suitable for any seed in which the embryo can be bisected. Most local chemists, and the educational suppliers listed in the back of this book, will be able to supply the Tetrazolium salt compound called '2,3,5-triphenyltetrazolium chloride'. Example embryo staining characteristics are given in Figure 2.2.

Aim

To test germination potential of seeds using the tetrazolium test.

Apparatus

graph paper
Tetrazolium salt
20 wheat and pea seeds

A-grade stain
(normal germination)

B-grade stain
(less than normal
germination)

C-grade stain
(poor germination)

D-grade stain
(very poor germination)

E-grade stain
(inability to
germinate)

Figure 2.2
Staining regions and germination potential in monocotyledon seed embryos

23

Method

1 Prepare a 1 per cent Tetrazolium salt solution (i.e. 1 g of salt mixed with 100 ml of water) in accordance with the manufacturer's instructions.

(a) Soak seeds overnight in water.
(b) Soak in 1 per cent triphenyl tetrazolium chloride solution, and store in darkness for 4 hours.
(c) Drain off tetrazolium solution, rinse with water and examine seeds within 2 hours.

2 Cut open 20 seeds of each species and observe the red stained embryos.
3 Observe the extent of the embryo staining and place each seed into a germination glass.

Results

1 Categorize each examined seed into grade A–E as indicated in Figure 2.2, entering your results in the tables provided.

Extent of embryo staining	Wheat seed Implied germination potential	Number in grade class	Percentage of total
A-grade	normal germination percentage		
B-grade	less than normal germination percentage		
C-grade	poor germination percentage		
D-grade	very poor germination percentage		
E-grade	inability to very poor germination		

Extent of embryo staining	Pea seed Implied germination potential	Number in grade class	Percentage of total
A-grade	normal germination percentage		
B-grade	less than normal germination percentage		
C-grade	poor germination percentage		
D-grade	very poor germination percentage		
E-grade	inability to very poor germination		

2 Produce a bar chart summarizing your results on graph paper.

Conclusions

1 Is this test for viability or vigour?
2 At what concentration is Tetrazolium salt mixed in distilled water?
3 How fast can germination results be achieved using this method?
4 Calculate a total germination percentage for the sample investigated.

Exercise 2.5 *Germination environment*

Background

All seeds are tested across a wide range of environmental conditions for their ability to germinate. This exercise investigates some of these properties.

Aim

To assess the effect of different environments on seed germination

Apparatus

 calculator
 filter paper
 lettuce and oil seed rape seed
 petri dishes

Method

1 Place a disc of filter paper in each petri dish and damp it down with water.
2 Sow petri dishes A, B and C with 10 seeds of oil seed rape and store in 5°C (fridge), 20°C (room temperature) and 30°C (incubator or airing cupboard) respectively.
3 Sow petri dishes D, E and F with 10 seeds of lettuce and store in 5°C, 20°C and 30°C respectively.
4 Monitor and keep the filter papers moist over the next 7 days.
5 After 7 days count the number of each seed germinated and measure the length of the plumule.

Results

Enter your results in the tables provided.

Germination environment oil seed rape	Total seeds sown	Number germin- ated	% germin- ated	Average plumule length
5°C				
20°C				
30°C				

Germination environment lettuce	Total seeds sown	Number germin- ated	% germin- ated	Average plumule length
5°C				
20°C				
30°C				

Conclusions

1 Is this test for viability or vigour?
2 Explain how temperature may affect the rate of seed germination.
3 What action can a professional horticulturalist take to increase the germination rate of seeds?
4 Design a suitable experiment further to test vigour ability under different environmental conditions, stating clearly your framework.

3 Propagation

Background

The reproduction of plants is called propagation. It may be either **sexual**, reproducing from seed, or **asexual** reproduction from existing plant tissue (**vegetative propagation**). The most difficult plants to propagate are also the most expensive.

Sexual propagation

Sexual propagation is mainly used for plant species where the variation of characteristics (e.g. size, colour, texture) is not important. Unlike dicotyledons, monocotyledons do not have the new growth producing cells called the cambium layer in the stem. The cambium cells are meristem (growth) points from which new roots would emerge. Since it would therefore be very difficult to take cuttings from monocotyledons, reproduction from seed is indispensable. Sexual methods are therefore essential for the propagation of **annual** plants (which live for only one season) and **biennial** plants (which are planted one year and flower the next before dying). To improve plant characteristics, for example in relation to resistance to disease, plants are bred by crossing with other stock to produce a superior cultivar (sexual propagation). This is called hybridization and involves raising the seed from cross-pollinated plants. New and improved plants can therefore only be created by raising seed, but once produced their population size may be rapidly increased through vegetative propagation techniques thus maintaining varieties and supply.

Vegetative propagation

Many plants, including trees and shrubs, are able to vegetatively propagate themselves rather than reproduce by seed production.

Other plants may require artificial methods (e.g. taking cuttings) and need great care if they are to be successful. Almost all plants sold – perennials, bulbs, corms, trees and shrubs – are vegetatively propagated. This is not only for convenience but because most plants are hybrids which will not breed true from seed. When seeds of these plants are sown they will produce a mixture of size, shape and colour features. Vegetative propagation is used for plants where identical plants to the mother are required and variation in characteristics is not desired. These are called **clones** because they are genetically the same plants.

Plant organs and propagation

The plant is composed of four primary organs (roots, stems, leaves and flowers), all of which can be utilized as material for propagation. Many plant organ modifications exist to enable natural vegetative propagation. Of these, the stem is worthy of special note. The most significant aspect of stems lies in their modifications as organs associated with asexual reproduction. Such vegetative propagation results from the growth of a bud on a stem. The bud produces a completely new plant with roots, stems and leaves. All daughter plants produced are identical 'clones' of the mother plant and may also serve as food stores (e.g. tubers). These enable a quick burst of growth in the spring using stored energy (e.g. iris rhizomes).

The following structures are modified stems of great importance in vegetative propagation and some will be considered in the exercises in this chapter.

Tubers	swollen underground stems, e.g. potato, dahlia, knot grass, glory lily, caladium
Rhizomes	underground stems, e.g. iris, rhus, ginger, Convallaria
Stolons	overground stems, e.g. strawberry runners, some grasses, Ajuga
Corms	compressed stems, e.g. crocus, cyclamen, gladiola
True bulbs	very short shoots, e.g. Narcissus, hyacinths
Buds	condensed shoots ready to explode into growth, e.g. Brussels sprouts and bulbils in Lilium

There are several methods used to increase plant numbers. These include cuttings, layering, division, budding and grafting. Of these, **cuttings**, **layering** and **division** will be investigated in the following excercises.

Exercise 3.1 *Vegetative propagation by cuttings*

Background

Taking cuttings should result in the production of new plants from pieces of stem, root or leaves of older plants. The daughter plant resulting from a cutting is a **clone** of the mother. There are several methods but the primary aim is to encourage the cut tissue to produce first roots and subsequently a shoot. If new shoots grow away before the root system has developed there will be an inadequate supply of water and the plant will die. Many of the methods used therefore concentrate on ways to increase the environmental conditions under which a vigorous root system will develop, e.g. applying basal root zone warming (bottom heat) and using **hormone** treatments.

Example methods include hardwood cuttings (rooting increased under mist thus keeping shoots short), softwood cuttings (root in peat, sand, soil and water), leaf cuttings (e.g. begonias and saintpaulias) and root cuttings (an occasional technique, e.g. horse-radish and hollyhocks). Suitable plants to take cuttings from include shrubs, conifers, many herbaceous and alpines (but not monocotyledons), some trees, some roses and climbers.

Good hygiene is essential to achieve superior results. Always use disease-free plants, clean equipment and fresh compost.

Using rooting hormones

The principal aim is to increase the percentage of cuttings that 'take' and therefore grow vigorously resulting in a healthy daughter plant. The benefits are:

● stimulation of root initiation
● a larger percentage of cuttings form roots
● faster rooting time.

Growth regulators stimulate the rooting of easy to root species. Hard, or failer to root species, normally show no response to hormone treatment.

Cuttings and rooting hormones

The hormone called auxin (IBA, Indole-butyric acid), is the most common rooting stimulant. This is because it is very persistent (long lasting) due to weak auxin activity, and effective as a root promoter. Because IBA translocates (moves around the plant) poorly, it is retained near the site of application. Hormones that readily translocate may cause undesirable growth effects in the propagated plant.

NAA (Naphthylacetic acid) is also very proficient at root promotion. However it is more toxic than IBA, and excessive concentrations are likely to damage or kill the plant.

The naturally occurring hormone is called IAA (Indole acetic acid). Both IBA and NAA are more effective at promoting roots than IAA, which is very unstable in plants. Decomposition of IAA occurs rapidly, unlike artificial compounds.

Growth regulators may alter the type of roots formed as well as the quantity grown. IBA produces a strong fibrous root system whereas NAA often produces a bushy, but stunted root system.

Root-promoting compounds work better when used in combination with others. 50 per cent IBA plus 50 per cent NAA results in a larger percentage of cuttings rooting than when either material used alone.

Method of application

There are three methods that work particularly well for stem cuttings.

Quick dip

The basal end of the cutting is dipped into a concentrated solution (500 to 10 000 parts per million, ppm) of the chemical dissolved in alcohol. Cuttings are dipped for varying lengths of time ranging from five minutes to six hours, depending on

(a) the species
(b) type of cutting
(c) age of tissue
(d) concentration of solution.

Solutions of auxin can be stored for fairly long periods of time provided they are in a closed container, out of direct light and chilled.

Prolonged dip

A concentrated stock solution is prepared in alcohol. This is then diluted in water at the time of use to give strengths of 20–200 ppm. Impurities in the water rapidly help to decompose the auxin and render it useless. The cuttings are soaked in this solution for up to twenty-four hours. Uptake of auxin tends to be more erratic and depends on the environmental conditions. It is therefore inconsistent.

Powder dip

The growth hormone is contained in an inert powder (e.g. clay or talc), at concentrations of 200–1000 ppm (softwood cuttings) or 1000–5000 ppm (hardwood cuttings). The cutting is dipped in the rooting powder and then planted. The auxin is then absorbed from the powder.

Aim

To assess the effectiveness of different hormones on the percentage 'take' of cuttings.

Apparatus

compost
geranium stock plants
IAA solution (100 ppm)
IBA solution (100 ppm)
50 per cent IBA and NAA solution (100 ppm)
NAA solution (100 ppm)
plant labels
scalpel
seed tray
small beakers
Seredix powdered rooting hormone (100 ppm) or alternative, e.g. Murphy Hormone rooting powder (captan + 1-naphthylacetic acid)

Method

1 Prepare the hormone solutions to the required concentrations, by dilution with distilled water.
2 Select three of the five hormone treatments for study.
3 Prepare three stem cuttings of the Geranium plant. You may collect your own additional cuttings (e.g. woody species) for investigation if you wish.
4 Label each cutting with your initials, date and hormone treatment under investigation.
5 Using the 'Quick dip' technique previously described, treat your cutting and place into a compost seed tray applicable for that treatment.
6 Record weekly the effectiveness of the treatments in the table provided.
7 On your next visit to the garden centre, record the type of hormones used in amateur rooting compounds.

Results

Enter your results in the table provided.

Cutting treatment at 100 ppm	Rooting response			Total cuttings treated	% successful take
	Week 1	Week 2	Week 3		
IBA					
NAA					
IAA					
50% IBA and NAA					
Seredix powder					

Conclusions

1 Why is an inert powder used?
2 Why do hardwood cuttings require a more concentrated hormone application?
3 Why is distilled water used?
4 Why is the daughter plant identical to the parent?
5 What is meant by breeding true?
6 What benefits are there from hybridization?
7 State the naturally occurring plant hormone.
8 What effect can an overapplication of NAA have on cuttings?
9 What advantages do artificial compounds have over the naturally occurring plant hormones?
10 Calculate the percentage of cuttings that formed roots.
11 State what the initials 'ppm' mean.
12 What other techniques can you state that increase the effectiveness of cuttings to root?

Exercise 3.2 *Propagation and layering (stem rhizomes and stolons)*

Background

Plant stems are often suitable propagation material to increase plant stock. Strawberry plants, for example, are often supplied in this form as 'runners'. In addition many weeds owe their success to an ability to evade some herbicides by reproducing this way (see also Exercise 3.3).

Rhizomes are underground horizontal stems growing from a lateral bud near the stem base. At each node there is a bud capable of sending out roots and shoots (e.g. iris, lily-of-the-valley, sand sedge, couch grass, canna lily, ginger, zandeschia, Solomon's seal, rhus and bulrush).

Stolons are overground stems. The stems bend over and form new plants at their tips (offsets). They have very long internodes with small scale leaves at the nodes. There is a bud at each node which can produce both roots and shoots. Eventually the stolon dies leaving an independent daughter plant (e.g. brambles, ground ivy, creeping buttercup, houseleek, strawberry and blackberry).

Layering

Stolon-forming plants naturally layer themselves (e.g. rhododendron and honeysuckle). New plants are formed when stems make contact with the surrounding soil where a new root system develops. This technique is suitable for most woody plants and shrubs. Stems are bent over and staked to the surface of the ground, thus encouraging the plant to root and shoot while still attached to the mother plant. The stolon can then be severed and the new plant grown in situ or potted-on (e.g. climbers, carnations, strawberries, jasmine).

Aim

To distinguish between rhizome and stolon stems as agents of vegetative propagation.

Apparatus

    iris and couch rhizomes
    strawberry runners

Method

Observe the specimens.

Results

Draw a labelled diagram of the couch rhizome and strawberry stolon, including the following terms if relevant: node, internodes, shoots, stolon, rhizome, mother plant, adventitious roots.

Conclusions

1 From where did the 'cloned' daughter plant form?
2 Define a rhizome.
3 Define a stolon.

4 Plants that have rhizomes are often resistant to flame weed killers. Explain why this is so.
5 On BBC Radio Four's *Gardener's Question Time* Mrs Davies asked why pulling up couch grass was not effective in eliminating the weed from her garden. What would be your answer if you were on the panel?

Exercise 3.3 *Propagation and division (stem corms and bulbs)*

Background

Corms and bulbs are further examples of modified stems and have a particular ability to perennate (survive over winter), thus giving good growth early in the next season.

Corms are underground vertically compressed stems which bear buds in the axils of the scale-like remains of the previous seasons leaves (e.g. crocus, gladiola, cyclamen, anemone, freesia and cuckoo pint). These are similar to bulbs but it is the stem that swells rather than the leaf bases.

True bulbs are underground swollen fleshly leaved, storage bases encircling a very short shoot (e.g. narcissus, lily, tulip, nerine, hyacinth, onion, snow drop and bluebell). A bulb contains the whole plant including the flower head in immature form. In spring the rapidly growing terminal bud uses the stored food in the swollen leaves to produce a flowering stem and leaves. Energy manufactured from photosynthesis is stored in the leaf base, which then swells forming a new bulb to perennate the next winter.

Division

This involves separating the root stocks to produce two new plants. This method is only possible with bulbs, herbaceous perennials and plants producing rhizomes or stolons. The growth medium should be thoroughly wetted-up first. Remove the mother plant and shake off excess soil or compost. Then divide the root ball either by hand, two back-to-back forks or a sharp knife, depending on the toughness of the root ball.

Best results are achieved when the plant is dormant during the winter months. Suitable material includes herbaceous and alpine plants. This technique is particularly important commercially for the propagation of raspberries and strawberries.

Division and plant organs

Many monocotyledons plants, and others that have fibrous root systems, can generate new plant material simply by dividing the plant at the root clump. Plants that reproduce by rhizomes and stolons can also be divided provided the stem contains a bud (e.g. *Stachys, Aster* and some ornamental grasses). The divided material is then simply potted-up, often taking the opportunity to cut back some foliage at the same time.

Suitable plants include herbaceous perennials (*Aster, Hosta, Stachys*), perennial herbs (chives and mint), alpine and rock plants (*Aubretia, Dianthus, Phlox, Sedum*), house plants (*Acorus, Aspidistra, Caladium, Chlorophytum, Cyperus, Saintpaulia, Scirpus* and stag's horn ferns) and succulents (*Aloe, Agave, Echeveria*).

For best results house plants should be divided in spring before new growth starts. Early flowering species should be divided in mid-summer immediately after flowering. Late flowering species should be divided between October and March. (See also Exercise 4.3.)

Aim

Investigation of bulbs and corms as tissue for propagation by division.

Apparatus

    crocus corms
    daffodil bulbs
    dissection kits

Method

1 Bisect the specimens.
2 Observe the corms' axillary buds and swollen stems.
3 Observe the bulbs' folded scales (leaves), flowers and stems.

Results

Draw a labelled diagram of your observations, showing clearly how corms differ from bulbs, including the following terms if relevant: flower bud, swollen leaves, future leaves, last year's leaf base, axillary bud, stem, adventitious roots.

Conclusions

1 Corms, tubers and rhizomes are often wrongly described as bulbs. Explain the differences.
2 Explain the advantages and disadvantages of corms and bulbs to plants in perennation.
3 What features make bulbs suitable as material for vegetative propagation?
4 Why are corms not suitable as material for vegetative propagation?

Exercise 3.4 *Propagation and division (bulbs)*

Aim

To utilize previously dissected bulbs from Exercise 3.3, to produce vegetatively propagated offspring using the 'Chipping' method.

Apparatus

    dissected bulbs (e.g. Narcissus)
    9:1:1 vermiculite, peat, water media
    0.2 per cent Benlate fungicide solution
    incubator 23°C
    scalpel

Method

1 Trim of the nose and stem of the dissected bulb.
2 Cut the bulb longitudinally into quarters.

3 Cut off slices from each quarter about 4 mm thick on the outside tapering into the centre.
4 Place slices (chips) into 0.2 per cent Benlate fungicide solution for 30 minutes.
5 Place chip in media tray, recording how many are inserted.
6 Incubate for 2 months at 23°C.
7 After 2 months small bulbs will have developed ready for planting out.

Results

1 Record the total number of incubated chips produced.
2 After 2 months record the number of successful bulbs.
3 Calculate the percentage success rate.

Conclusions

1 Explain the variation in results between species.
2 Why do you think some chips were unsuccessful?
3 How might the percentage success rate be improved?
4 List other methods of vegetative propagation with which you are familiar.

4 Weed – biology and control

Background

A weed is any undesirable plant hindering or suppressing the growth of more valued plants through competition for water, nutrients or light and may in addition harbour insects carrying viruses or disease. Weeds tend to invade every area of horticulture from sports turf, amenity plantings, nursery stock to glasshouse soils. The effects can be devastating, especially if they bring other pests and disease with them. They have aspects to their biology that make them successful, such as fast germination rates or quantity of seed produced. In addition, control methods may vary at different growth stages. Annual weeds in particular are susceptible to contact herbicide control at the seedling stage, before the cotyledons have dropped off, but not at the adult stage. Weed identification can be a difficult skill to master and for this reason it is worthwhile starting with weeds in flower as this aids speedy identification. Weeds are normally divided into annual and perennial weeds.

Annual weeds

These complete their life cycle, from germination to death, in one season (e.g. chickweed, groundsel and speedwell). They are quick growing throughout the year and produce abundant seeds which can remain dormant in the soil for several years.

Perennial weeds

These plants normally flower annually and live for more than two years (e.g. creeping thistle, couch, yarrow and broad-leaved dock). They often die back during the winter and regrow when the temperature rises in spring. They are characterized by tap roots and rhizomes that regrow if chopped into small pieces such as when rotavating.

Exercise 4.1 *Weed collecting and identification*

Aim

To recognize and collect some annual and perennial weeds in the adult and seedling stages, from different horticultural situations.

Apparatus

> hand lens
> microscopes
> sample bags
> trowels
> wildflower handbook, e.g.
> Fitter R.S.R. (1985) *Wild Flowers of Britain and Northern Europe*, Collins
> Phillips, R. (1986) *The Photographic Guide to Identify Garden and Field Weeds*, Elm Tree

Method

1 Using the wild flower handbooks as a guide, collect three different annual and three different perennial weeds from one, or more, of the following horticultural situations:

  (a) in a lawn
  (b) in a flower border/bed or in a growing field crop
  (c) below established trees/shrubs
  (d) in a hard landscaped area, e.g. paving or gravelled area
  (e) in a glasshouse soil.

The following weeds are suggested targets for collection:

Annuals	Perennials
chickweed (*Stellaria media*)	creeping thistle (*Cirsum arvense*)
groundsel (*Senecio vulgaris*)	couch grass (*Agropyron repens*)
speedwells (*Veronica* spp.)	yarrow (*Achillea millifolium*)

2 Also try collecting seedling specimens, with their cotyledons present, to see at which stage they could be controlled.
3 Use the trowel with care to dig up as much of the plant as possible including the roots.
4 Place in sample bag for later observations.
5 Observe the specimen under the microscope or using a hand lens noting any significant features.
6 Using the wild flower handbook, identify and name each weed and state the characteristics of the weed that make it successful (e.g. annual, habit of plant, acid soil lover).
7 Record the location and site features of where the weed was found (e.g. soil type, aspects of location and how this differs from site to site).
8 Save the specimens for use with subsequent exercises.

Results

Enter your results in the table provided.

Horticultural situation	Name of weed	Botanical characteristics that make the weed successful	Date found	Location found	Details of site
Lawn	1				
	2				
	3				
Flower border/field crop	1				
	2				
	3				
Established trees or shrubs	1				
	2				
	3				
Hard landscaped area	1				
	2				
	3				
Glasshouse soil	1				
	2				
	3				

Conclusions

1 State what is meant by the term 'weed'.
2 List four features that make annual weeds successful in cultivated areas.
3 Give an example of an annual and perennial weed.
4 List four harmful effects of weeds.
5 Were there any associations between the location of the weeds that you found and the surrounding community/landscape? If so, what?

Exercise 4.2 *Pressing and mounting weed specimens*

Background

Once competence has been gained at weed recognition it will be useful to keep a record to aid future quick reference and identification. This can be achieved by pressing the plant material in a plant press or under a flat heavy object, between sheets of paper to absorb the sap. The press flattens and dries plant material. When dry they can be mounted on white card and secured with a little glue. They may then be protected by covering with 'Transpaseal' or other form of sticky-back plastic. However, the pressing stage is the most critical and if not properly done, mould and other fungal growths will appear under the transpaseal, concealing any chance of later identifying the specimen. Thick and succulent materials are particularly vulnerable, but this can be easily avoided by changing the paper daily at first, then less often once most of the sap has been removed.

Aim

To produce a collection of pressed weeds for future quick reference and identification.

Apparatus

absorbent paper
plant press
weeds specimens (from Exercise 4.1)

Method

Follow the instructions in Figure 4.1.

Figure 4.1

Pressing and mounting weed specimens

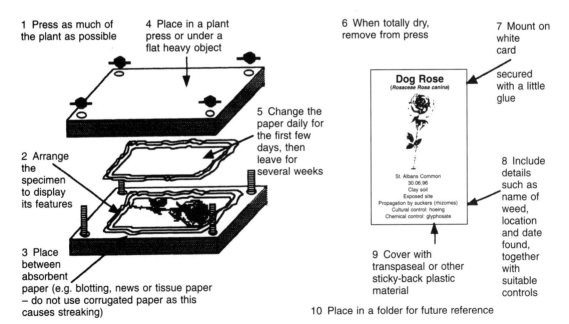

1 Press as much of the plant as possible

4 Place in a plant press or under a flat heavy object

6 When totally dry, remove from press

7 Mount on white card

secured with a little glue

5 Change the paper daily for the first few days, then leave for several weeks

Dog Rose
(Rosaceae Rosa canina)

St. Albans Common
30.06.96
Clay soil
Exposed site
Propagation by suckers (rhizomes)
Cultural control: hoeing
Chemical control: glyphosate

2 Arrange the specimen to display its features

3 Place between absorbent paper (e.g. blotting, news or tissue paper – do not use corrugated paper as this causes streaking)

8 Include details such as name of weed, location and date found, together with suitable controls

9 Cover with transpaseal or other sticky-back plastic material

10 Place in a folder for future reference

Results

Depending on the species drying may take several weeks to several months.

Conclusions

1 How should the material be arranged on the press?
2 What causes the build-up of mould on some pressed specimens?
3 Explain what difficulties pressing succulent plants presents.

Exercise 4.3 *Roots structure*

Background

The root system of both weeds and cultivated plants plays an important role in vegetatively propagating new plants, particularly following cultivations when the root may be chopped into several pieces each capable of producing a new plant. This exercise is designed to gain familiarization with the structure of plant roots, not only for weeds but also for general horticultural practices. This knowledge is developed in subsequent exercises.

The roots are the primary organ for the uptake of water into the plant. But they have other functions such as:

(a) to anchor the plant to the ground
(b) to absorb water and dissolved mineral salts from the soil
(c) to act as a pipeline between the soil and the stem
(d) modified, as organs of food storage and vegetative propagation.

Most roots are either tap roots (mainly dicotyledons) or fibrous roots (mainly monocotyledons).

Tap Root

These have a strong main root descending vertically with little or no lateral growth. These roots are often strengthened by lignin and become very hard to remove and inflexible. Conifer tap roots can penetrate chalk 20–30 m deep, providing good support and resistance to drought. Some tap roots when chopped up produce new plants (e.g. chrysanthemum and docks) and therefore make weed control very difficult. Some trees and herbaceous perennials are vegetatively propagated from root cuttings (e.g. many perennials and biennials, dandelion, dock, parsnip and carrot) (see also Propagation exercises on Division).

Fibrous root system

A root which branches in all directions in a mass of fine matted roots rather than thick, fleshy ones. These explore the top soil much more effectively than tap roots. However as they are shallow rooting, they suffer in dry weather and need irrigation. This can be difficult since water passes through the soil profile quickly and passes beyond the reach of the roots. These plants are very susceptible to drought and require accurate watering if plant loss is to be avoided (e.g. most annuals, groundsel, grass, ferns and house plants in general).

Adventitious roots

These occur in unusual or unexpected locations, such as the negatively phototropic roots on the side of juvenile aerial stems of ivy. New roots which form from stem cuttings are also called adventitious roots. Iris rhizomes shows adventitious roots at their nodes.

Rooting depth

In summer conditions, when the soil is drying out, water may not move by capillary action fast enough towards the root system to keep pace with transpiration loss. Roots must continually grow into new damp areas of soil to maintain the water supply. Soil compaction and pans could prevent this. Roots consequently have a large surface area to improve water absorption, but are disadvantaged by being susceptible to diseases (e.g. club root).

Container-grown plant roots

Woody plants grown in containers need regular potting-on into larger containers so that roots have space to grow otherwise they become pot-bound – roots tend to spiral round the pot and plants may die, even up to twelve years after subsequent planting out.

Aim

To investigate how different root structures influence the success and management of plants.

Apparatus

adventitious roots of ivy
binocular microscope
fibrous grass roots
tap root from broad leaved dock

Annual weeds	**Perennial weeds**
chickweed (*Stellaria media*)	creeping thistle (*Cirsum arvense*)
groundsel (*Senecio vulgaris*)	couch grass (*Agropyron repens*)
speedwells (*Veronica* spp.)	yarrow (*Achillea millifolium*)

Method

1 Observe the specimen tap and fibrous roots under the microscope.
2 Observe fine root hairs under the binocular microscope.
3 Observe and classify the annual and perennial weed's roots into 'tap' or 'fibrous' classes.

Results

Enter your results in the table provided.

Life cycle type	Weed species	Root structure
Annual weeds	chickweed (*Stellaria media*)	
	groundsel (*Senecio vulgaris*)	
	speedwells (*Veronica* spp.)	
Perennial weeds	creeping thistle (*Cirsum arvense*)	
	couch grass (*Agropyron repens*)	
	yarrow (*Achillea millifolium*)	

2 Root hairs allow water to enter the root by osmosis (see Chapter 7). Estimate the length of root hairs on a 1 cm length of your specimens.
3 Draw a labelled diagram of the parts of tap and fibrous systems including, main root, lateral root, root hairs, where appropriate.

Conclusions

1 Which root system would be more susceptible to drought conditions and why?
2 Which root system would be least susceptible to draught conditions and why?
3 Explain what precautions should be taken when growing woody plants in containers.
4 How can the roots system help you to understand how plant species become successfully adapted to their environment?
5 How would you tell which root system was the most lignified?
6 How can monocotyledons and dicotyledon species mostly be separated by their root systems?
7 List four features that make perennial weeds successful.

Exercise 4.4 *Cultural weed control*

Background

Generally, weed control methods are either 'cultural' or 'chemical'. Chemical control will be considered in Exercise 4.5. Cultural control refers to cultivation techniques of 'good practice' – tending carefully the growing plants, and removing competitors, without using chemicals. It includes the use of mechanical cultivators and rotavators. There are three main cultural control methods.

Hoeing   Hoeing simply involves bringing the weed to the surface where the action of sun and the absence of water kills the plant. For this reason it is more effective in hot, dry weather. This is particularly suited for annual weeds, provided the roots are fully exposed. Hoeing and rotavating always bring more weeds to the surface through disturbing the soil.

Rotavating   In addition to turning the soil this practice chops weeds up into small pieces. However, perennials, are more resilient, largely because of their ability to regenerate from root cuttings and, for effective control, this practice must be continued for the whole season until the action of rotavation has exhausted the plant's food supply. This feature will be investigated in this exercise.

Flame   This treatment has tended to be restricted to glasshouse crops
weed killers   where some degree of soil sterilization is also required. Small ride-on machines, with their own gas supply, are now available. Steam sterilization is also a common glasshouse practice.

Other techniques include; hand weeding with a small fork; improving soil structure, especially drainage (e.g. spiking and aeration); creating a stale seed bed by irrigating to encourage annual weeds to germinate which are then killed before replanting; purchasing certified clean seed that has been screened against weed seeds; raking, so that stems are brought upwards to meet blades of hoe (e.g. to remove clover *Trifolium repens*); scarifying to remove dead organic matter preventing build-up of a microclimate that favours certain weeds (e.g. moulds and annual meadow grass); reducing shade (e.g. to discourage lesser celandine *Ranunculus ficaria* from lawns); and mulching to prevent light reaching weeds.

Aim

To investigate the effectiveness of different rotavation techniques to prevent the regrowth of annual and perennial weeds.

Apparatus

secateurs
soil filled seed trays (×7)

Annual weeds
chickweed (*Stellaria media*)
groundsel (*Senecio vulgaris*)
speedwells (*Veronica* spp.)

Perennial weeds
creeping thistle (*Cirsum arvense*)
couch grass (*Agropyron repens*)
yarrow (*Achillea millifolium*)
broad-leaved dock (*Rumex obtusifolius*)

Method

1 Collect examples of the annual and perennial weeds listed above.
2 Cut the roots of each plant into lengths of 1 cm, 2 cm, 5 cm, 7 cm and 10 cm lengths.
3 Record the number of cuttings made in each size band.
4 Plant cuttings in a labelled seed tray.
5 Leave for 7 days to regrow.
6 After 7 days record the number of cuttings in each size band that have successfully regrown.

Results

Enter your results in the table provided.

Type of Weed		Cutting size-band	Number of cuttings planted	Number regrown after 7 days	Percentage regrowth	Percentage not regrown
Annuals	chickweed (*Stellaria media*)	1 cm				
		2 cm				
		5 cm				
		7 cm				
		10 cm				
	groundsel (*Senecio vulgaris*)	1 cm				
		2 cm				
		5 cm				
		7 cm				
		10 cm				
	speedwells (*Veronica* spp.)	1 cm				
		2 cm				
		5 cm				
		7 cm				
		10 cm				
Perennials	couch grass (*Agropyron repens*)	1 cm				
		2 cm				
		5 cm				
		7 cm				
		10 cm				
	creeping thistle (*Cirsum arvense*)	1 cm				
		2 cm				
		5 cm				
		7 cm				
		10 cm				
	yarrow (*Achillea millifolium*)	1 cm				
		2 cm				
		5 cm				
		7 cm				
		10 cm				
	broad-leaved dock (*Rumex obtusifolius*)	1 cm				
		2 cm				
		5 cm				
		7 cm				
		10 cm				

Conclusions

1 Why are cultivation techniques unsuitable for eliminating dock weeds?
2 State the rotavated cutting size necessary to prevent regrowth for each of the investigated weeds.
3 How effective was rotavation at eliminating the growth of annual weeds?
4 How effective was rotavation at eliminating the growth of perennial weeds?
5 Describe how the grower can affect the growth of weeds by cultivation practices.
6 State four methods of controlling weeds without using herbicides.

Exercise 4.5 *Chemical weed control*

Chemical control

Do not use weedkillers indiscriminately. Several are toxic to beneficial insects and can do ecological damage. Herbicide use is now covered by the **Food and Environmental Protection Act** (FEPA) of 1985, and requires operators to be certified as competent users. Resistance may build up if chemicals are persistently used and some degree of crop rotation should be practised to reduce this risk. *The UK Pesticide Guide* (Whitehead, 1995), is an essential reference source to identify suitable herbicides for use in horticulture and is reviewed annually. The use of herbicides has greatly reduced labour costs and also reduced the wide row spacing previously necessary between some crops (e.g. carrots).

Chemical weed control normally involves the use of some form of modified plant hormone as a growth regulator. For example, the use of 2,4D (a translocated herbicide) is a form of synthetic auxin. When sprayed on dicotyledonous weeds (e.g. thistles) it causes abnormal growth. The inter nodes become severely elongated, the phloem becomes blocked and the meristem points produce mutated cells. The root system becomes corrupted and death follows swiftly thereafter.

Chemical weed controls may be classified as:

- non-selective – kill all plants whether weeds or crops
- selective – kill weeds but do not harm crops. This is called '**differential toxicity**'. Different groups of plants have different responses to the same concentration of weed killer, e.g. dicotyledons are more sensitive than monocotyledons to 2,4D, thus when sprayed on to a grassed lawn (monocotyledons), thistle weeds (dicotyledons) die off. Similarly with Propyzamide herbicide. Some examples of the concentration (in parts per million) required to cause plant death, given to demonstrate differential toxicity are: crop – carrot (0.8 ppm), cabbage (1.0 ppm), lettuce (78.0 ppm); weeds – knot grass (0.08 ppm), fat hen (0.2 ppm), groundsel (78.0 ppm)
- contact – kill only those parts that they touch
- translocated – absorbed and travel through the phloem to kill the meristem points in both roots and shoots. This method is essential to kill the root system of large established weeds
- non-residual – act immediately after application, e.g. paraquat
- residual – remain active in the soil for long periods of time
- foliage acting – absorbed through the leaves
- soil acting – absorbed through the roots.

Some herbicides have different characteristics with varying crops. Some herbicides for example, are absorbed through both leaves and roots; others are selective in one situation and non-selective in another. The roots and shoots of the same plant will respond differently to the same concentration of weed killers. In fact many herbicides contain a mixture of the various different substances.

Aim

To select suitable herbicides to control annual and perennial weeds at different growth stages, in a range of horticultural situations.

Apparatus

The UK Pesticide Guide (`the green book'; Whitehead, 1995).

Method

Use the current edition of the *UK Pesticide Guide* to find suitable herbicide details to control the following annual and perennial weeds, at both the seedling and adult stage.

Results

Enter your results in the table provided.

Horticultural situation and name of weed	Growth stage	Active ingredient	Product name	Manufac-turing company	Precautions to be taken when using this product
Lawn yarrow (*Achillea millifolium*)	Seedling				
	Adult				
Flower border/field crop speedwells (*Veronica* spp.)	Seedling				
	Adult				
Established trees or shrubs broad-leaved dock (*Rumex obtusifolius*)	Seedling				
	Adult				
Hard landscaped area groundsel (*Senecio vulgaris*)	Seedling				
	Adult				
Glasshouse soil creeping thistle (*Cirsum arvense*)	Seedling				
	Adult				

Conclusions

1 List four types of herbicides available to control some common weeds.
2 Name one translocated and one contact herbicide.
3 State two ways in which selectivity of herbicide takes place.
4 State two conditions necessary for the successful use of each of the following herbicides:

(a) contact
(b) translocated
(c) residual.

5 What legislation governs the use of herbicides in the UK?

5 The leaf and photosynthesis

Background

Photosynthesis is the process enabling plants to make food, in the form of carbohydrates (sugars and starch), using light energy. Sugars are manufactured in the leaf from carbon dioxide (CO_2) and water (H_2O), using the energy of sunlight trapped by chlorophyll and other pigments. Oxygen (O_2), is the waste product given off by the plant. The sugars are transported around the plant (via the phloem) and deposited as starch. For most plants photosynthesis can occur over the temperature range 5–30°C. The optimum is considered to be 20–25°C, depending on species.

The rate of photosynthesis is influenced by the environment and as horticulturalists we can regulate the rate of photosynthesis by altering the supply of these factors. These include:

- Leaf area and age   Managed by de-leafing, stimulating new growth
- Chlorophyll distribution   Managed by de-leafing, stimulating new growth
- Light   Managed by supplementary lighting, blackouts, shading
- Carbon dioxide levels   Managed by CO_2 enrichment
- Temperature   Managed by heating, ventilation
- Water supply   Managed by irrigation, drainage
- Nutrients   Managed by fertilizers, liquid feeding.

The **Law of the Limiting Factor** states that the factor in the least supply will limit the rate of process. We cannot increase photosynthesis by increasing a factor already in adequate supply. We must first identify which is the lowest factor and increase this. For example we cannot increase carbon dioxide levels continually and expect to see an increase in growth without increasing other photosynthesis factors such as temperature as well.

The leaf is the main site for photosynthesis. The larger the leaf surface area, the faster will be the rate of photosynthesis and the subsequent growth and development of the plant. The leaf is one of the four primary organs of the plant. It may be defined as a lateral organ to the stem or axis of a plant below its growing point.

The leaf has four main functions, most of which relate to the growth and development of plants through the processes of photosynthesis:

- manufacturing food by photosynthesis
- enabling the diffusion of gases from leaf to atmosphere
- transpiration and cooling
- with modifications, as organs of food storage and vegetative propagation.

If we can therefore understand the structure and workings of the leaf the horticulturalist can increase the growth rate and ease of establishment of plants.

Different plant species have become adapted and specialized to their natural habitats around the world. Plant organs have evolved to most efficiently survive that environment and, even in a garden situation, the correct positioning of plants to take advantage of these features is important.

Some leaves are adapted to aid **support**, e.g. sweet pea tendril which is sensitive to touch (haptotropism) and twists around objects to support climbing. Plants that live in water are called **hydrophytes** and have extra air spaces in their cells to trap oxygen (e.g. water lily, pondweed, duck weed), and subsequently flop when removed from their environment. By comparison **succulent** leaves are fleshy with their own water supply (e.g. echeveria, kalanchoe). **Carnivorous** plants are suitable for sites with poor nutrition and contain leaves modified to trap and digest insects (e.g. Dionaea (venus fly trap), sundew, butterwort and pitcher plants).

Starch test procedure

If photosynthesis is taking place the leaves should be producing sugars. Sugars are soluble materials that can be transported through the plant and converted into insoluble materials called starch. However, in many leaves, as fast as sugar is produced it is turned into starch. Since it is easier to test for the presence of starch than sugar, we can regard the production of starch as evidence that photosynthesis has taken place. The word and chemical formulas for the photosynthetic reaction are:

$$\text{Carbon dioxide} + \text{Water} \xrightarrow[\text{Chlorophyll}]{\text{Light energy}} \text{Sugar} + \text{Oxygen}$$

or

$$6CO_2 + 6H_2O \xrightarrow[\text{Chlorophyll}]{\text{Light energy}} C_6H_{12}O_6 + 6O_2$$

Thus, carbon dioxide and water are changed by chlorophyll using light energy into sugar and oxygen. The following exercises are designed to test this theory.

Aim

To test for the activity of photosynthesis (follow this procedure in Exercises 5.2–5.5).

Apparatus

Starch testing apparatus tray containing:

test tube rack	iodine
boiling tube (100 ml)	matches
pyrex beaker	Bunsen burner
gauze mat	tripod
forceps	dropping pipette
pencil	industrial methylated spirits (IMS)
2 stirring rods	white tile

CAUTION

Please note the following hazards

During these exercises you will be working with industrial methylated spirit (IMS). IMS is **highly flammable**. Do not use near an open flame. IMS has an **irritating vapour**. Do not inhale. If so, move into the fresh air. IMS is **harmful to the skin and eyes**. Wear eye goggles and take care not to spill any. If so, wash with plenty of water under a running tap.

Method

You may find it helpful to draw a labelled diagram of the apparatus.

1 Three quarters fill the pyrex beaker with water and place on the gauze mat, on a tripod over the Bunsen burner.

2 Light the Bunsen burner and heat the water until it boils.

3 Remove a leaf from the plant and boil in the water for 2 minutes. This ruptures the cell walls and prevents any further chemical reactions occurring.

4 Put on safety goggles and, using the forceps, remove the leaf and place into the test tube of IMS.

5 Turn off the gas to the Bunsen burner.

6 Place the test tube into the water bath and allow to warm until the leaf turns completely white (usually 5–10 minutes). This indicates that all the chlorophyll has been removed and makes the reaction of iodine with starch more clearly visible. The leaf will now be brittle and hard. IMS is dangerous and not to be heated directly. Always ensure that the tube is pointing away from you.

7 To return the leaf softness dip it again into the warm water bath.

8 Place the leaf onto a white tile and drench with iodine.

9 Observe the reaction over the next 2 minutes and record your results in the tables provided with each exercise.

Results

Positive: The reaction of starch (white), in the presence of iodine (yellow), is to change the leaf colour to deep blue or purple/black.

Negative: If no starch is present then the iodine will stain the leaf yellow or brown.

Exercise 5.1 *Dicotyledonous leaf structure and photosynthesis*

Background

What is it about leaves that make them so important for photosynthesis, and growth and development?

In typical leaf structure diagrams adaptations to specific environments are described through leaf studies (e.g. margins, shapes and apexes). The amazon palm, for example, has a leaf blade 20 m long! The British broom, by comparison has tiny leaves. Leaves may be either broad (typical of dicotyledons) or narrow (typical of monocotyledons). They often have distinctive textures such as small hairs which trap air to keep the plant cool (e.g. *Stachys lanata*). Similarly, *Erica carnea*, *Calluna vulgaris* and rosemary are adapted to dry conditions and save water by curling their leaves creating a humid atmosphere.

There are several layers of different types of leaf tissue. These layers all have specialist jobs to perform to maximize the growth and development of the plant. Leaves have a **large surface area** onto which sunlight can fall. A waterproof waxy layer of **cuticle** covers the epidermis and is thicker on the upper surface of the leaf giving it a shiny appearance (e.g. camellia). Each time leaves are polished a layer of cuticle is removed and house plants are then more susceptible to wilting and disease.

The **epidermis** contains tightly fitting cells which helps to keep the leaf air-tight, protecting the leaf from water loss and from bacteria and fungal penetration which might cause diseases. The epidermal layer is transparent allowing light to pass through to the mesophyll layers below. This layer contains the chlorophyll and is arranged into two layers of tissue. First, the **palisade layer**. This is a layer of closely packed cells containing the green pigment called **chlorophyll**. Chlorophyll produces the green colour in plants and are the sites were photosynthesis (the manufacture of sugar) actually takes place. Second, the **spongy mesophyll**. This layer of loosely packed irregular shaped cells has many air spaces allowing gaseous exchange between the leaf and the atmosphere. This enables carbon dioxide to get to the palisade layer, and allows waste gases and moisture to escape through the stomata.

The veins (called vascular bundles) of leaves contain xylem and phloem transport vessels. Water and minerals are supplied to the leaf through the xylem. The phloem vessel transports the manufactured sugars, formed during photosynthesis, away from the leaf.

The **lower epidermis** contains small pore-like openings called stomata. These are pores in the underside of leaves. They are the entry and exit points for the diffusion of gases and water. The opening and closing of the stomata is controlled by the water content of the surrounding guard cells.

Aim

To relate the functions of leaves to their component parts.

Apparatus

> binocular microscopes
> transitional section dicotyledon leaf slide

Method

Observe the specimen slides under the microscope.

Results

Draw a labelled diagram including the following terms:

air spaces	palisade layer	upper epidermis
lateral veins	spongy mesophyll	vascular bundle
lower epidermis	stomata	waxy cuticle
main vein		

Conclusions

1 State the function of plant leaves.
2 Explain how you would select a plant suitable for a shallow, dry soil.
3 What leaf features would make a plant suitable for a sunny, exposed location?
4 Where is the leaf tissue found that is responsible for photosynthesis?

5 Explain what is meant by the term chlorophyll.
6 How do leaves reduce the risk of fungi and bacterial infection?
7 Name two of the tissues in the leaf that play a role in photosynthesis and state their particular function.
8 State four factors which effect the rate of photosynthesis.
9 Describe, with the aid of a diagram, how the structure of a dicotyledon leaf is designed to facilitate photosynthesis.
10 Explain the 'Law of the limiting factor'.

Exercise 5.2 *Chlorophyll and photosynthesis*

Background

Chloroplasts contain chlorophyll. This is the main plant pigment able to capture light energy and use it to turn carbon dioxide and water into sugar and oxygen. It is a mixture of pigments including chlorophyll-a (blue/green leaf colours), chlorophyll-b (yellow/green colours), carotene (orange leaf colours), xanthophyll (yellow autumn colours) and authocyanne (red/purple leaf colours). Chlorophyll is mainly found in the palisade layer of leaves. Variegated plants contain chlorophyll only in their green parts, only these areas should contain starch. However, in deserts the leaves of cacti are reduced to spines. The stem contains chlorophyll and are able to photosynthesize.

Chlorophyll production is partly dependent on nutrient supply and can be greatly influenced by our management of soil pH and fertilizer use. For example, a high soil pH (through over liming) causes iron and manganese deficiency. This creates yellowing on the youngest leaves due to an absence of chlorophyll formation. In addition, over fertilizing with potassium causes magnesium deficiency resulting in yellowing of the middle and older leaves and an absence of chlorophyll production.

Aim

To test whether or not chlorophyll is necessary for photosynthesis by conducting a starch test on a variegated plant leaf and observing the locations of positive reactions.

Apparatus

    a variegated plant (e.g. geranium) labelled A
    starch-testing apparatus tray

Method

1 De-starch plants. At the beginning of each exercise, place the plants in darkness for a few days. In darkness the starch is changed into sugar and transported out of the leaf.
2 Remove plant A from darkness and place in sunlight for several hours before beginning.
3 Remove a leaf from plant A and draw a diagram of its shape and colours (green and white).
4 Conduct a starch test on the leaf as detailed in the chapter backgound text and record your results in the table provided.
5 Draw a second diagram to show the subsequent colour change on reaction with iodine.

Results

The reaction of starch (white) in the presence of iodine (yellow) is to change the colour to deep blue or black. If no starch is present then the iodine will stain the leaf yellow or brown.

Variegated leaf colour	Colour reaction with iodine	Interpretation – starch present or absent
Green		
White		

Conclusions

1 What is chlorophyll?
2 What conclusions can be made about chlorophyll and photosynthesis?
3 Do variegated plants have more or less starch reserves?
4 In which vessel is the sugar solution transported away?
5 Name the main pigment responsible for light absorption in photosynthesis.
6 Write out a word equation that summarizes the reactions in photosynthesis.

Exercise 5.3 *Light and photosynthesis*

Background

The formula for photosynthesis implies that light is necessary for photosynthesis to occur. We regard the production of starch as evidence that photosynthesis has taken place. If we increase either the amount of light or provide light of greater intensity (brightness), more photosynthesis can take place. There are several techniques that the professional horticulturalist may use to achieve this, including:

- artificial light
- use of reflective surfaces (white wash)
- shading
- blackouts
- orientation of glasshouse or garden to south facing
- increased plant spacings
- reduced plant spacing
- remove weeds/eliminate competition
- clean glasshouse panes

Aim

To assess if plants can photosynthesize in both light and dark conditions.

Apparatus

geranium plants labelled B and C
starch-testing apparatus

Method

1 De-starch plants B and C before the exercise (see Exercise 5.2 for procedure).
2 Place plant B in sunlight for several hours. Keep plant C in darkness.
3 Remove a leaf from both plants B and C.
4 Conduct a starch test as detailed in the chapter background text.

Results

The reaction of starch (white) in the presence of iodine (yellow) is to change the colour to deep blue. If no starch is present then the iodine will stain the leaf yellow or brown.

Light conditions	Colour reaction with iodine	Interpretation – starch present or absent
Plant B (light)		
Plant C (dark)		

Conclusions

1 Which plant leaf went blue in the presence of iodine?
2 What can you conclude about the leaf that was kept in darkness?
3 What can you conclude about photosynthesis and the requirement for light?
4 Explain how light influences plant growth and development.
5 Describe the effect of low or no light on plant growth.
6 Explain why plant growth rate is lower in cloudy weather.
7 Describe how light levels may be manipulated to increase the rate of photosynthesis.

Exercise 5.4 *Carbon dioxide and photosynthesis*

Background

Carbon dioxide (CO_2) is believed to be a raw material in the process of photosynthesis. Can plants photosynthesize in a carbon dioxide-free atmosphere?

Carbon dioxide comes from the air and enters the plant leaves through the stomata. The natural concentration of carbon dioxide in the air is 0.035 per cent (350 ppm). In unventilated glasshouses plants use up carbon dioxide and the subsequent reduction in supply causes the rate of photosynthesis to slow down. By carbon dioxide enrichment in glasshouses, this concentration can be raised three times to 0.1 per cent (1000 ppm), leading to an increase in the rate of photosynthesis and increased plant growth.

Enrichment often occurs during the winter months on sunny days when an enriched atmosphere can be maintained because the vents are closed. An injection rate of 15–55 kg/ha/hr (kilograms per hectare per hour), under moderate wind conditions is normal. In the summer months when crops are growing vigorously carbon dioxide levels in the crop may fall to 0.01 per cent (100 ppm) and even with fully open vents injection rates of 25–30 kg/ha/hr are required to maintain concentrations at normal atmosphere levels (0.035 per cent or 350 ppm). Economic returns vary with the size of glasshouse and other variables, but as a general rule for every 1p spent on carbon dioxide enrichment, a return of 4p is yielded through increased crop yield.

Aim

To sustain a plant in a carbon dioxide free environment and observe whether or not photosynthesis is possible by monitoring the presence of starch in the plant leaves.

Apparatus

delivery tubes × 5	lime
demi-john chamber	soda lime canister
geranium plant labelled D	starch testing apparatus
glass platform	water flasks × 2
high vacuum gel	water pump

Method

You may find it helpful to draw a labelled diagram of the apparatus.

1 De-starch plant D and seal into the carbon dioxide-free chamber for several hours (see Exercise 5.2 for procedure).
2 Remove a leaf from plant D.
3 Conduct a starch test as detailed in the chapter background text and record your results in the table provided.

Results

The reaction of starch (white) in the presence of iodine (yellow) is to change the colour to deep blue. If no starch is present then the iodine will stain the leaf yellow or brown.

CO_2 free environment	Colour reaction with iodine	Interpretation – starch present or absent
Plant D		

Conclusions

1 Why is soda lime used?
2 Explain why lime water flasks are used.
3 How does carbon dioxide enter the plant?
4 What can you conclude about carbon dioxide and photosynthesis?
5 Explain why a lack of carbon dioxide reduces photosynthesis.
6 Describe how carbon dioxide levels may be manipulated to increase the rate of photosynthesis.

Exercise 5.5 *Water and photosynthesis*

Background

Water is important to plants in many ways including maintaining leaf structure for photosynthesis. Water is supplied to the leaf through the xylem vessel. Although only 1 per cent of water absorbed by plants is used for photosynthesis, any lack of water in the leaves causes water stress. Water is essential to maintain leaf turgidity (cell strength), without which it would not be possible to open the stomata and allow carbon dioxide into the leaf. A plant that is wilting cannot therefore photosynthesize, and if allowed to continue to wilt would result in plant death.

Some plants are adapted to prevent water loss. For example in gorse, broom and horsetails the leaves are reduced to scales, lowering the surface area and reducing water loss. The stems contain chlorophyll and photosynthesize. There are several techniques that professional horticulturalists may use to prevent water loss. Here are some examples of both plant and horticulturalist strategies.

Plant strategies

- thick cuticles
- close stomata
- reduce surface area
- create a humid atmosphere, e.g. curl leaves

Horticulturalist's strategies

- shading
- increase humidity, e.g. fogging, mist, polythene
- lower the air temperature
- reduce leaf area
- reduce air movements
- spray with anti-transpirants

Aim

To observe if plants can photosynthesize in the absence of water.

Apparatus

> geranium plant labelled E
> starch test apparatus

Method

1 De-starch plant E before the exercise (see Exercise 5.2 for procedure).
2 Do not water plant E for several days before the exercise.
3 Remove a leaf from plant E.
4 Conduct a starch test as detailed in the chapter background text and record your results in the table provided.

Results

The reaction of starch (white) in the presence of iodine (yellow) is to change the colour to deep blue. If no starch is present then the iodine will stain the leaf yellow or brown.

Water-free environment	Colour reaction with iodine	Interpretation – starch present or absent
Plant E		

Conclusions

1 What can you conclude about water and photosynthesis?
2 How does the water enter the plant for photosynthesis?
3 How many units of water are required in photosynthesis?
4 List two ways in which plant leaves are adapted to reduce water loss.
5 Write out a chemical equation that summarizes the reactions in photosynthesis.

Exercise 5.6 *Oxygen and photosynthesis*

Background

Photosynthesis results in the production of sugar, which is stored as starch and used later for growth and development (respiration) and oxygen, which is a waste product.

Aim

To observe the production of oxygen by pond weed under light and dark conditions.

Apparatus

    beakers ×2
    glass funnel
    Griffin oxygen metre
    Plants labelled F and G (pond weed)
    test tube

Method

You may find it helpful to draw a labelled diagram of the apparatus.

1 Keep the plants in a beaker of water throughout the exercise, covered by a glass funnel.
2 De-starch plants F and G before the exercise (see Exercise 5.2 for procedure).
3 Keep plant F in darkness. Place plant G in the light for several hours.
4 Observe the gas being produced under each condition.
5 Record the oxygen content of the water using the oxygen metre.

Results

Enter your results in the table provided.

Light/dark environment	Oxygen level recorded	Interpretation
Plant F (light)		
Plant G (dark)		

Conclusions

1 What can you conclude about oxygen and photosynthesis?
2 How does the waste oxygen escape from the plant?
3 Complete the gaps in the following text:

For a plant to produce carbohydrates, it absorbs from the surrounding air
_____ and _____ in the form of gases. These enter the plant
through the _____ on the leaves. The carbohydrates produced by
the plant are used as _____ in the process of growth. Water is also
supplied through the _____ vessels within the leaf veins (vascular
bundles).

4 List the pair of ingredients of photosynthesis under the following headings:

Consumable inputs
Resulting products
Processes systems.

5 List five factors that affect the rate of photosynthesis.
6 List the methods a horticulturalist can use to maximize the rate of photosynthesis during the winter months.

6 Respiration and storage

Background

Respiration is the destruction (use) of sugars and starch, made during photosynthesis, to release energy for the plant's life sustaining processes. In aerobic conditions (with oxygen) the waste products are carbon dioxide gas and water vapour.

All organisms need to respire. Respiration is essential for plant growth and development. The liberated energy is used for cell division and making useful plant substances, including cellulose in cell walls, proteins, enzymes and general growth and repair.

The process of respiration takes place in all living cells in the plant. It occurs most rapidly at the meristematic points (growth points) where cells are actively dividing (e.g. roots and shoots). It is a continual process occurring during both day and night.

Because energy is used for plant growth and development, respiration results in a decrease in plant sugars. This is acceptable if the sugars are being replaced through photosynthesis, but in certain circumstances this is not possible and plants may subsequently shrivel and die, for example wilting plants, cut flowers, stored fruit and vegetables.

In these conditions attempts must be made to slow down respiration and extend the life of the plant. Respiration rates are influenced by a variety of environmental factors including:

- temperature
- water
- oxygen level
- plant injury (wounding or infection).

Respiration may be either aerobic (with air) or anaerobic (without air).

Aerobic respiration

Sugar + Oxygen $\rightarrow$ Carbon dioxide + Water + Energy

$C_6H_{12}O_6 + 6O_2 \rightarrow 6CO_2 + 6H_2O + 2830\,KJ$

Anaerobic respiration

Sugar $\rightarrow$ Carbon dioxide + Alcohol + Energy

$C_6H_{12}O_6 \rightarrow 2CO_2 + 2C_2H_5OH + 118\,KJ$

These exercises are designed to investigate these ingredients and to ascertain the best storage method to reduce respiration, and therefore extend product shelf life.

Exercise 6.1 *Storage of plant materials*

Background

While the plant is growing, respiration rates are needed that are high enough to provide the energy for the production of new structures, but not so high that excessive amounts of energy are lost in unnecessary vegetative growth. The grower must balance environmental factors such as temperature and light, with respiration, so that the sugars broken down by respiration do not exceed those produced by photosynthesis, otherwise there will be negative growth.

However, when plants are harvested for selling or display respiration continues. Plant sugars are being used up without being replaced, as they are not photosynthesized, and the produce, be they flowers, vegetables or fruit, will eventually break down and decompose. There are several circumstances where it is desirable to slow down the rate of respiration to prevent decay, such as:

Night time: Respiration occurs during both day and night, but photo-synthesis occurs only during the day. The plant must therefore store enough sugar during the day to fuel night-time respiration. Respiration increases with temperature. If we lower the temperature at night (by ventilation) then we can slow down the rate of respiration.

Storage: Respiration continues when plants are harvested. Therefore sugars are being used up without being replaced and the plant will eventually decay. We therefore seek to reduce respiration by putting the produce in cold stores (e.g. 0–10°C storage for cut flowers, apples, vegetables, onions, cuttings and chrysanthemums).

Similarly, respiration increases with oxygen levels. Therefore if we lower the oxygen supply in controlled-atmosphere stores decay will be reduced (e.g.

cut flowers stored in a florist's shop). However, if oxygen levels are very low, aerobic respiration stops and anaerobic respiration starts. Alcohol is a waste produced and is toxic to plants. In apple storage low oxygen levels, caused by carbon dioxide build-up, result in an alcohol flavour. In comparison, a high carbon dioxide level leads to brown heart in pears and core flush in apples.

Aim

To assess the effectiveness of different storage systems on reducing respiration rates in carrots and therefore extending the shelf life of our produce (cut flowers are also suitable for investigation).

Apparatus

5 carrots	plastic bags
top-pan analytical balance	sample labels
incubating ovens	indelible pens
cold storage fridge	

Method

1 Arrange for one carrot to be stored under all of the following environmental conditions:

   A   high temperature incubator (30°C), sealed in a plastic bag
   B   high temperature incubator (30°C), without a plastic bag
   C   cold storage fridge (4°C), sealed in a plastic bag
   D   cold storage fridge (4°C), without a plastic bag
   E   ambient (room) temperature without a plastic bag
   F   ambient (room) temperature in a plastic bag.

2 For those carrots that are placed in a plastic bag, please place a label on the carrot and weigh the carrot before placing in the bag.
3 After labelling, weigh your carrots before storing them using the analytical balances and the following technique:

   (a) 'Tare' (zero) the balance. This sets the microprocessor memory to read zero.
   (b) While the microprocessor is doing this, the balance readout display will be blank. Do not put anything on the balance while this is happening.
   (c) When the display returns to '0.00 g' commence weighing your sample.

4 Record the masses in the table provided.
5 Label your samples, using the indelible pens, providing the following information:

   (a) name
   (b) date sample prepared
   (c) weight of carrots recorded
   (d) storage method A, B, C, D, E, or F.

6 Leave your samples in storage for two weeks.
7 After two weeks:

   (a) remove your samples from storage and reweigh them
   (b) record this as the mass after storage, in the table provided.

Results

1 Calculate the difference in weight between before and after storage.
2 Calculate the percentage change in weight using the following formula:

$$\text{Percentage change in mass} = \frac{\text{mass after storage} - \text{mass before storage}}{\text{mass before storage}} \times 100$$

Storage treatment method	Mass before storage (g)	Mass after storage (g)	Change in mass (g)	% change in mass
A				
B				
C				
D				
E				
F				

Conclusions

1 Explain your results in terms of respiration rates and storage treatments.
2 Rank your storage treatments in order of their efficiency at reducing respiration.
3 Explain the significance of respiration rate in relation to storage of harvested produce and seeds.
4 State two instances where horticulturists might want to slow down the rate of respiration.
5 Explain two ways in which a reduction in the rate of respiration might be achieved.

63

Exercise 6.2 *Aerobic germination of pea seeds*

Background

Aerobic respiration is the process whereby carbohydrates (sugars and starch) are broken down using oxygen to liberate energy, with the production of waste carbon dioxide and water. Carbon dioxide gas, when mixed with lime water, will turn the water milky and is therefore an indicator that respiration has taken place. Without water respiration cannot take place (e.g. seeds will not germinate without water).

Aim

To demonstrate that germinating seeds respire and to identify the gas given-off.

Apparatus

    water-soaked pea seed
    water-soaked boiled peas
    cotton wool
    corked conical flasks A and B
    test tube of lime water

Method

You may find it helpful to draw a labelled diagram of the apparatus.

1 Place the living, water-soaked seeds, in conical flask A.
2 Place the boiled peas in flask B.
3 Leave both flasks several days to germinate.
4 Uncork both flasks and tip the gas into the lime water. Carbon dioxide is heavier than air and, although colourless, will flow if the flasks are poured gently into the lime water.

Results

1 Record what happens to the lime water in flasks A and B.
2 What is the gas being produced?
3 What process results in the production of this gas?
4 Why were some of the peas boiled?
5 What effect does boiling the seed have on germination ability?

Conclusions

1 What are your conclusions from this experiment?
2 Explain the meaning of the term 'aerobic respiration'?
3 State two factors which increase the rate of respiration.

Exercise 6.3 *Anaerobic respiration of yeast*

Background

Anaerobic respiration is respiration without oxygen. Alcohol is produced as a waste product and is toxic to plants. Anaerobic respiration produces less energy for the same amount of sugar than aerobic respiration (4 per cent) (from $118/2830 \times 100$).

$$\text{Glucose} \rightarrow \text{Carbon dioxide} + \text{Alcohol} + \text{Energy}$$

$$C_6H_{12}O_6 \rightarrow 2CO_2 + 2C_2H_5OH + 118\,KJ$$

There are several circumstances in horticulture where this occurs. Management action should be taken to prevent anaerobic conditions developing. Examples of where such environments occur include:

- poorly aerated soils – result in poor growth/germination
- poor film in hydroponic production – results in low oxygen and low growth
- modified atmosphere packaging – results in mould growth and alcohol flavour
- compaction – results in poor growth, hydrogen sulphide gas (rotten egg smell)
- overwatering – results in root death, stunted growth.

Aim

To demonstrate that anaerobic respiration can take place. This exercise is designed so that the yeast is living in an anaerobic environment.

Apparatus

delivery tube	live yeast
lime water filled test tube	test tube
liquid paraffin	water bath

Method

You may find it helpful to draw a labelled diagram of the apparatus.

1 Prepare de-oxygenated water by boiling distilled water.
2 To the test tube add 5 ml of glucose solution (5 per cent mass/volume or m/v), prepared using de-oxygenated water, and 1 ml of yeast suspension (10 per cent m/v), also prepared using de-oxygenated water.
3 Cover the mixture with a thin layer of liquid paraffin to exclude oxygen from the mixture.
4 Connect the test tube of lime water to the test tube of yeast, using a delivery tube.
5 Place the yeast mixture test tube in a warm water bath to stimulate growth.
6 Leave the mixture for 30 minutes and await a reaction.

Results

1 How long does it take before the lime water reacts?
2 What happens to the lime water?
3 What is the gas given off?
4 What process results in the production of this gas?
5 Why is distilled water used?
6 Why is deoxygenated water used?

Conclusions

1 Explain what is meant by 'anaerobic respiration'.
2 State two places where anaerobic respiration might occur in horticulture and state its importance.
3 Explain why anaerobic conditions are undesirable in horticulture.

Exercise 6.4 *Energy release during respiration*

Background

Heat being produced is a good indication that energy is being used up during respiration. Respiration is a chemical reaction; chemical reactions increase with temperature, therefore respiration increases with temperature. For every 10°C rise in temperature between 10°C and 30°C, the respiration rate will double. At temperatures above 30°C, excessive respiration occurs, which is harmful to the plant. At temperatures above 40°C, plant tissue collapses and breaks down.

Respiration rates may also be stimulated when the plant has been injured in some way. Under these conditions the plant starts to heal itself through producing new cells to repair the damage. This new growth is fuelled by respiration. Any plant injury to tissue, either by mechanical damage or by infection, increases the rate of respiration. For example, damaged stems caused by overcrowding in a bucket of cut flowers results in sugars being used up as the plant repairs the damage.

However, there are times when we want to increase plant respiration rates to increase growth and development. The respiration rate at the base of the cuttings is increased after the cutting is taken by applying heat to the root zone (basal root zone warming). Similarly seed respiration rates increase as they germinate. They need energy to break down cotyledon food store and produce new cell growth.

Aim

To investigate heat as a product of respiration during the germination of wheat seeds.

Apparatus

vacuum flasks labelled A and B
cotton wool
1 litre of wheat seeds
thermometers labelled A and B
1 per cent formalin solution

Method

You may find it helpful to draw a labelled diagram of the apparatus.

1 Rinse the wheat seeds in 1 per cent formalin solution, to kill all bacteria and fungi on the grains
2 Place half of the wheat seeds in vacuum flask A.
3 Place half of the wheat seeds in vacuum flask B.
4 Place a laboratory thermometer in both flasks A and B.
5 Seal each flask with cotton wool and leave to germinate for a few days.
6 After a few days record the temperature of flasks A and B, in the table provided.

Results

Enter your results in the table provided.

Seed treatment	Recorded temperature °C
Flask A	
Flask B	
Temperature difference	

Conclusions

1 Which flask has the highest temperature?
2 Why are the flasks sealed with cotton wool?
3 What can be concluded about respiration from this experiment?
4 Describe how the natural healing ability of plants is exploited in plant propagation.
5 Explain the significance to the grower of the process of plant respiration in obtaining optimum temperatures for growth of a crop.
6 Compare the process of respiration with photosynthesis.

7 Plant water

Background

Between 70 and 95 per cent of plant matter is water (by fresh weight). A lack of water is probably the most important factor in the loss of crop yield. Water in plants has a variety of purposes, including:

- constituent of cell sap
- participant in a number of chemical reactions
- photosynthesis
- maintains turgidity of cells
- transpiration
- source of protons (hydrogen ions)
- solvent for vital reactions
- medium through which substances move from cell to cell
- medium for the transportation of substances around the plant.

The water content of plants may vary from 70–95 per cent in growing plants, to as low as 5–10 per cent in dormant structures such as seeds. Excluding water, a plant is made up of organic material synthesized from carbon dioxide, water, nitrogen and the mineral matter absorbed through the roots. Once absorbed, minerals remain in the plant and are only lost through leaf fall. Water, however, is constantly being lost. This effects the rate of absorption and movement of minerals. In order to photosynthesize plants must get carbon dioxide into the leaf. This is encouraged by transpiration as the plant exposes leaves to the air. This water loss means additional water must be taken up to replace it.

The absorption and transportation of water occurs as the result of several different processes which will be investigated in the following exercises including:

(a) water movement into plant roots – osmosis, plasmolysis, diffusion
(b) water movement within the plant – root pressure osmosis, capillary rise.

The following exercises will investigate these processes:

1 model of how osmosis occurs
2 diffusion
3 osmosis
4 plasmolysis
5 osmosis, diffusion and plasmolysis determination
6 root pressure osmosis.

Exercise 7.1 *Model of osmosis*

Background

Osmosis is the movement of water from a dilute (weak) solution across a semipermeable membrane, to a more concentrated (strong) solution (e.g. the smaller scale movement of water from cell to cell).

A semipermeable membrane (e.g. plant root), is a sieve of tiny pores too small to allow large molecules like sugar ($C_6H_{12}O_6$) to pass through, but large enough to let small water molecules (H_2O) through.

Osmosis is the method by which water is passed into and out of cells. The cells gain or lose water until the solutions are equal. It is therefore important in many functions such as the opening and closing of the stomata and plant turgidity. Air-dried seeds, starches, proteins and cellulose attract water. Water is therefore described as imbibed as water moves from an area of high water content to an area of lower water content. Some people like to think of osmosis as the ability of the plant to suck water into itself.

As water moves across the membrane into the cell, it leads to a build-up of pressure, this develops cell turgidity (turgor). Turgidity enables stems to stand upright and leaves to be held firmly to the light. Normally living cells have a slightly lower than maximum turgor pressure, therefore they can absorb water by osmosis. If water is lost from cells they become flaccid (wilt).

Aim

To predict the movement of solutions of different strengths.

Method

Observe Figures 7.1 and 7.2 and state which way the water will flow (osmotic direction pull). NB. All molecules are in constant motion. The membrane is permeable only to water. The sucrose molecules are too large to pass through.

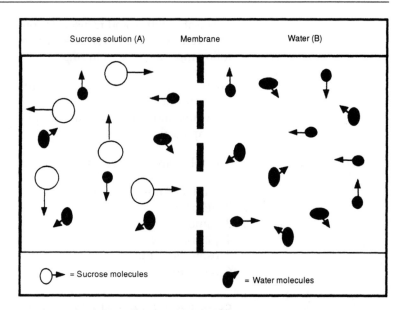

Figure 7.1
Osmotic direction pull for solutions A and B

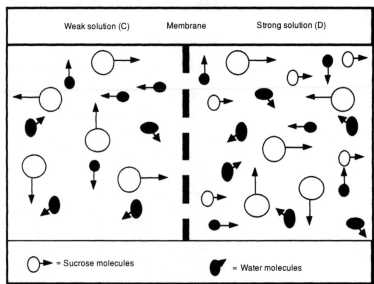

Figure 7.2
Osmotic direction pull for solutions C and D

Results

With reference to Figure 7.1:

1 What will happen to the two solutions?
2 Which way does the water flow?
3 Which two parts of the plant might the membrane represent?

In Figure 7.2:

4 What will happen to the two solutions?
5 Which way does the water flow?

6 What happens to plant cells when osmosis moves water into them?
7 What would happen if the strong solution was the soil water and the weak solution was the root sap?

Conclusions

1 State, in your own words, what is meant by osmosis.
2 Explain why dried raisins swell when placed in water.
3 Explain how, without drying you could cause the raisins to shrivel.
4 Explain what would happen to a vase of cut flowers if the owner added more than the recommended dose of Baby Bio plant food to the vase water.
5 How do plants take up water through their roots?

Exercise 7.2 *Diffusion*

Background

Diffusion is the movement of gas, liquids or salts from regions of high concentrations to low concentrations until the concentrations are equal (e.g. hot and cold water mixed to make warm water).

Diffusion occurs because molecules are in constant motion trying to produce conditions where they are evenly distributed. It is a slow process, where the rate of diffusion depends on the concentration gradient. This is the difference between high and low concentrations. The greater the difference, the faster will be the diffusion. For example, plants transpire water faster through the leaves on a hot sunny day than on a wet damp day when the diffusion gradient between the moist air outside the leaf and moist air inside the leaf is much narrower. Similarly, if we add fertilizers to the soil, we create a zone of high concentration relative to that surrounding the roots and, once they are in solution (e.g. after rainfall or irrigation), nutrients move towards the roots by diffusion. Some plant processes that utilize diffusion include:

- water vapour transpired through stomata
- nutrients move through the soil towards roots
- movement of dissolved nutrients between cells
- water movement through the soil to plant roots
- gas movement (CO_2 and O_2) into and out of leaves.

Aim

To demonstrate the process of diffusion.

Apparatus

water
500 ml beaker
potassium permanganate crystals ($KMNO_4$)

Method

You may find it helpful to draw a labelled diagram of the apparatus.

1 Fill the beaker with deionized water to the 500 ml mark.
2 Slowly add a few crystals of potassium permanganate.
3 Record the reaction.

Results

1 Give a written description of your observations.
2 How long did it take for the crystals to diffuse fully into the water?

Conclusions

1 Are the following statement true or false?

    (a) 'Water vapour moves by diffusion through the stomata, from an area of high concentration inside the leaf, into the air, an area of low concentration outside the leaf.'
    (b) 'Carbon dioxide moves from the air, an area of high concentration, through the stomata and into the leaf, an area of low concentration, by diffusion.'

2 How does a high atmospheric humidity affect diffusion of water?
3 Explain how the rate of transpiration is affected by:

    (a) spraying pot plants on display
    (b) exposure to wind or drafts
    (c) mist propagation.

4 What is the effect of covering display plants with a damp cloth?
5 Write, in your own words, a simple definition of 'diffusion'.
6 Explain how water moves through the soil to plant roots.

Exercise 7.3 *Osmosis*

Background

Osmosis is the movement of water from an area of low salt concentration, across a semipermeable membrane, into an area of higher salt concentration. It is important in many stages of the water cycle from root to atmosphere including the following.

Entry into the root hairs. Each root hair is a long cell containing a large central area of sap in which various salts and sugars are dissolved. The soil particles are surrounded by a film of water which, although containing salts, is a weak solution compared with that of the cell sap. They are separated by the living cell wall, which acts as a semipermeable membrane. Consequently, osmosis will occur and water enters root hair cells.

Passage across the root. The absorption of water into a root hair cell dilutes its contents. There are now two cells of different concentrations next to one another and the root hair cell acts as a relatively weak solution in comparison with the cell internal to it. Water therefore passes by osmosis to the internal cells, which in turn become diluted and so pass water into the cells internal to them. A complete osmotic gradient is established, extending from the root hair cells to the central cells of the root. Water passes along this gradient until it enters the cells around the xylem.

Aim

To demonstrate the process of osmosis.

Apparatus

2 potatoes, A and B
2 saucers
copper sulphate crystals

Method

You may find it helpful to draw a labelled diagram of the apparatus.

1 Boil Potato A for 10 minutes.
2 Hollow out each potato and place on a saucer.
3 Fill each saucer with water.
4 Add a few crystals of copper sulphate to each hollow.
5 Observe the reaction.

Results

Enter your observations in the table provided.

Sample	Hollow description	Saucer description
Potato A		
Potato B		

Conclusions

1 Explain why copper sulphate crystals were used in this exercise.
2 For potato B, state which area is the zone of high concentration (hollow or saucer).
3 Why is potato A boiled?
4 Name the process occurring.
5 What would happen if copper sulphate was placed in the saucer and water in the hollows?
6 Describe the process of osmosis and its importance in water movement through plants.
7 How does water move across the plant roots to the xylem vessel?
8 Explain the process involved that enables seed to absorb water prior to germination.
9 How are stomata guard cells able to open and close?
10 State four functions indicating the importance of water in plants.
11 Explain the relationship between water and plant growth.

Exercise 7.4 *Plasmolysis*

Background

> Plasmolysis is excessive water loss from the plant cell causing the protoplasm to shrink away from the cell wall.

Sometimes the soil solution can have a higher nutrient salt content than the root sap and osmosis will occur as water moves out from the plant (low salt level) into the soil (high salt level). The plant cells collapse (become flaccid), resulting in temporary wilting. When placed in a solution of lower concentration (e.g. water), turgor returns as water moves back into the cells. If left too long in the plasmolysis state permanent wilting occurs and the plant will die.

Plasmolysis often occurs when seeds are left to germinate in a recently fertilized soil, such as when creating a grass lawn. The seed will dry out, wither and die. If container-grown plants are left to dry out for too long between watering, the transpiration from the plant and evaporation from the container will cause an increase in the salt concentration of the solution and plasmolysis may occur. In the alkali flats of the USA, evaporation exceeds rainfall. Salts accumulate in the soil and plasmolysis follows. Water is drawn out from the plants and normal plants cannot grow. Plants especially adapted to survive these conditions are called halophytes.

Aim

To demonstrate the process of plasmolysis in horticultural situations.

Apparatus

> concentrated fertilizer solution (e.g. Tomorite tomato feed)
> deionized water
> two pots of germinating oil seed rape seedlings, A and B

Method

You may find it helpful to draw a labelled diagram of the apparatus.

1 Drench rape pot A with deionized water.
2 Drench rape pot B with concentrated fertilizer solution.
3 Record the reaction over the next hour.

Results

Enter your results in the table provided.

Sample type	Description of treatment method result	
	Deionized water	*Concentrated fertilizer*
Rape pot A		
Rape pot B		

Conclusions

1 Name the process that occurred

2 State what is meant by plasmolysis.

3 What might cause plasmolysis in a horticultural situation and state one method of rectification.

4 Explain why newly transplanted seedlings often wilt.

5 Describe what happens to root cells that are in contact with a soil solution with a high salts level.

6 Explain the impact of dog urine , or 'bitch scorch' (which is extremely rich in potassium ions), on lawn growth.

Exercise 7.5 *Osmosis, diffusion and plasmolysis*

Background

Now that you have investigated the processes of osmosis, diffusion and plasmolysis your skills will be developed to enable you to distinguish each process accurately. In this exercise a semipermeable membrane is used to simulate a plant root placed in different sucrose solutions. A dye is used to aid your observations.

Aim

To distinguish between osmosis, diffusion and plasmolysis.

Apparatus

3 × 500 ml beakers
red dye
deionized water
3 retort stands and clamps
3 dialysis semipermeable membranes
20 per cent sodium chloride solution (NaCl)

Method

1 Suspend each membrane from the retort stand clamp into a beaker.

2 Prepare the membrane and beaker solutions as follows:

(a) Beaker A: red dye + water in membrane, water in beaker
(b) Beaker B: red dye + 20 per cent NaCl solution in membrane, water in beaker
(c) Beaker C: red dye + water in membrane, 20 per cent NaCl in beaker.

3 Record the reaction when the membrane is placed in the beaker, noting particularly, any changes in water levels.

Results

You may find it useful to draw a diagram showing the apparatus before and after treatment. Enter your results in the table provided.

Situation	Change in water level		Interpretation of process occurring
	Beaker	Membrane	
Beaker A			
Beaker B			
Beaker C			

Conclusions

1 State the process occurring in beaker A.
2 State the process occurring in beaker B.
3 What would happen to plant cells subjected to the conditions that occurred in the membrane of beaker B?
4 State the process occurring in beaker C.
5 What would happen to plant cells subjected to the conditions that occurred in the membrane of beaker C?
6 Describe how water moves:

   (a) through the soil to plant roots
   (b) across the plant roots to the xylem
   (c) up the plant stem to the leaves.

7 Explain what happens to root cells which are in contact with a soil solution with a high concentration of fertilizer salts.

Exercise 7.6 *Root pressure osmosis*

Background

Water is aided to move up a plant by root pressure osmosis. This causes water to move through the root and into the xylem vessel of the stem from where it will travel to the rest of the plant. It is induced by the root pumping ions into the xylem.

The cells of the xylem are specialized and empty. They are arranged in columns, one on top of another, so that they form a system of hollow tubes through which water can be passed up the plant. Root pressure osmosis causes water to be passed into the xylem vessel. The water follows passively by osmosis and builds up in the xylem under pressure, normally only two to three atmospheres, occasionally seven or eight bar (one bar approximately equals one atmospheric pressure). Each bar pressure can support about 10 m of water. Root pressure is the main force left to move ions and water up the plant during the night and during the dormant season.

In North America the giant redwoods (*Sequoia*) can be up to 100 m tall. This means that the column of water that must be held in the xylem against the force of gravity is equivalent to the pressure required to lift a weight of 1.6 tonnes resting on the palm of your hand. Part of the force required to sustain such conditions is produced by the osmotic root pressure. This creates a pressure gradient in the xylem and water continues to move upwards through a process called capillary rise.

Aim

To demonstrate root pressure osmosis.

Apparatus

capillary tube
geranium pot plant
red dye

rubber sleeve
saucer
water

Method

1 Crop the geranium at the level of the compost.
2 Attach a capillary tube to the stem via the rubber sleeve.
3 Fill the capillary tube with red dye to 5 cm and draw a line over the miniscus.
4 Water the plant via the saucer for 10 days.
5 Record the change in the capillary water level after 10 days.

Results

You may find it helpful to draw a labelled diagram of the apparatus. Note how many millimetres the water level had risen in the capillary tubing.

Conclusions

1 To which of the following was the result due:

(a) photosynthesis
(b) osmosis
(c) root pressure osmosis
(d) diffusion
(e) plasmolysis.

2 Explain the importance of root pressure osmosis to plant growth and development.

8 Water transportation pathways and processes

Background

Water in the plant may be considered as one interconnected plumbing network. The lower region is in the soil. Root hairs are in intimate contact with soil water molecules. The upper region is separated from the atmosphere by a thin leaf layer. Plant water is therefore one giant water molecule held together by hydrogen bonds, so loss or absorption at one point affects the whole system.

How do the processes concerned with plant water operate in practice to move water through the soil into the plant and out into the atmosphere? There are several stages, including:

1 passage through the soil to the plant roots by diffusion (see Chapter 7)
2 entry into the root hairs by osmosis (see Chapter 7)
3 passage across the root and into the xylem by osmosis (see Chapter 7)
4 passage up the stem to the leaves by:

    (a) root pressure osmosis. This enables water to move through the root and into the xylem vessel of the stem from where it will travel to the rest of the plant (see Chapter 7)
    (b) capillary rise. As water travels up the plant the xylem vessels narrow creating further upwards pressure
    (c) transpiration pull/stream. The loss of water by transpiration from the leaves will therefore draw water, by osmosis, from the xylem, which in turn draws further water upwards from the roots

5 passage out of the plant into the atmosphere by (a) transpiration and (b) evaporation.

Some of these topics have been covered in the previous chapter. The main focus of this chapter will be on:

- stems as organs for passage of water up to the leaves
- leaves as organs enabling transpiration into the air
- transpiration in a range of horticultural environments
- methods for horticulturalists to reduce water loss from plants.

Exercise 8.1 *Stem tissue functions*

Background

The stem is the major organ through which water and nutrients are transported around the plant. It contains both unspecialized tissue, such as parenchemya packing/support cells, and specialized tissue, such as vascular bundles. Within the vascular bundles of dicotyledonous plants, three types of specialized tissue are found (xylem, phloem, cambium). The xylem vessel is strengthened by a compound called lignin to enable it to withstand the tremendous pressures that build up in transporting the weight of water upwards. The phloem is strengthened with cellulose and transports the sugars made during photosynthesis around the plant to where it is needed. The cambium, which is absent in monocotyledonous plants, contains specialized growth cells (meristems) and is responsible for the manufacture of new xylem and phloem cells. In addition new roots grow from the cambium after cuttings have been taken.

Aim

Investigation into why plants need stems and the purposes of the tissue that stems contain.

Apparatus

List of stem tissues including:

cambium	epidermis	phloem
cell membrane	fibres	vascular bundles
cytoplasm	parenchyma	xylem
endodermis		

Method

Match each stem tissue to the function indicated in the sentences below.

Results

1 Regulates movement of substances in and out of the cell.
2 Protein and water.

3 Vessel for the transport of water up the plant.
4 Vessel for the transport of nutrients around the plant.
5 The growing tips of roots and shoots, small cells producing new xylem and phloem vessels.
6 Cell's protective sheath.
7 Regulates water uptake in the roots.
8 Groups of strands of conducting tissue composed of xylem, cambium and phloem.
9 Unspecialized soft tissue packing cells.
10 Thread-like cells or filaments.

Conclusions

1 Explain why it is more difficult to take cuttings from monocotyledons than dicotyledons.
2 Describe how xylem tissue is adapted for water transportation around the plant.

Exercise 8.2 *Transport through plant stems*

Background

One of the plant's four primary organs is the stem. Why do plants need stems and what functions do they perform?

A stem may be defined as the leaf, bud or flower-bearing axis of the plant for support and transport. Sometimes a stem is wrongly called a stalk. A stalk is the common term used for the tissue supporting a flower or leaf. The correct term for a leaf stalk is petiole and for a flower stalk is pedicel. A stem can produce buds, a stalk cannot.

The stem is the main part of the plant above the ground. The functions are:

(a) to hold leaves to the light
(b) to support the flower in suitable positions for pollination
(c) as a pipeline between the roots and leaves, supplying water and nutrients for photosynthesis
(d) as a pipeline between the leaves and roots, distributing sugars manufactured by photosynthesis
(e) modified, as organs of food storage and vegetative propagation
(f) in green stems to manufacture food by photosynthesis.

Most stem tips (meristems) grow vertically towards the light (positively phototropic) and away from the force of gravity (negatively geotropic). If the tip is removed (pinched out) growth is prevented (e.g. gladioli to prevent stem twisting). Once it starts growing the plant normally has only one main stem (apical dominance, e.g. tomatoes). This main stem will grow straight upwards rather than in a bushy fashion. In some plants that do not exhibit apical dominance it is necessary to pinch out side buds preventing shoot growth from these areas and maintaining growth upwards.

Stems have become modified to take advantage of specific environments. Stems of timber, for example, are strengthened to counteract the resistance that the mass of leaves offers to the wind. Stem shapes are also important in supporting the plant such as the square stem of a nettle, or the five-ridged stem of a wall flower. Other plants with naturally contorted stems include twisted willow (*Salix matsudana* 'Tortuosa') and contorted hazel (*Corylus avellana* 'Contorta'). In dry conditions stems may act as water storage organs (e.g. succulents, cacti and baobab tree).

There are several differences between stems of monocotyledonous and dicotyledonous specimens including the content and arrangement of vascular bundles, which help to explain the modifications described above and also influence the passage of water through the plant. This exercise will investigate these differences.

Aim

To investigate differences between monocotyledon and dicotyledon stem structures as the plumbing system of the plant.

Apparatus

    monocotyledon stem slide
    dicotyledon stem slide
    monocular microscope
    chrysanthemum
    lily

Method

1 Cut open a stem of chrysanthemum and lily and compare the different tissues seen. These are similar to the tissue which will be enlarged on the slide under the microscope. How does the arrangement of the vascular bundles differ between them?
2 Observe each specimen slide under the microscope.

Results

Draw fully labelled diagrams to show the distribution of tissues in the primary stem of a named dicotyledonous and monocotyledonous plant. Include the following tissues if present:

    cambium   parenchyma
    cell membrane   phloem
    epidermis   vascular bundles
    fibres   xylem

Conclusions

1 State two functions of stems.
2 Name two types of cell involved in transporting water and nutrients around the plant.
3 Describe the difference in monocotyledonous and dicotyledonous plant stems with reference to:

(a) the arrangement of the vascular bundles
(b) the tissue content of the vascular bundles .

4 State the function of the following parts in the dicotyledonous stem:

(a) xylem
(b) cambium.

81

5 State the functions of the following stem tissues:

(a) phloem
(b) epidermis.

6 Where, in a flowering plant, does cell division take place?
7 Where are parenchyma cells found in plant stems?
8 Describe, with the aid of diagrams, the differences between the internal structure of a dicotyledonous and monocotyledonous stem, and relate this to their functions.

Exercise 8.3 *Water loss from leaf stomata*

Background

Leaves are punctured by air spaces, communicating with the atmosphere by the stomata. The stomata are tiny openings in the epidermis of leaves and allow the entry and exit of gases (mainly carbon dioxide), including moisture vapour, between the leaf and the atmosphere. Through these openings water is able to escape from the plant during transpiration.

Such **gaseous exchange** is essential for photosynthesis and respiration life systems. This occurs through a processes called **diffusion** (see Chapter 7). The cells surrounding air spaces in the spongy mesophyll contain moisture, which will evaporate into the air spaces and diffuse into the atmosphere when the stomata are open.

The stomata is the main site for **transpiration** without which water and minerals would not be able to move around the plant. Transpiration also aids leaf cooling as the moisture evaporates. For water molecules to change from liquid to gas (vapour) form, they need energy to break their hydrogen bonds. At 20°C this takes 2450 J for every 1 g of water. Consequently, the evaporation of water by transpiration also helps cool the leaf. For example, on a dry, bright summer day, the energy extracted from the leaves of a large rose bush during transpiration is approximately 250 watts.

The opening and closing of the stomata (singular stoma) is controlled by the amount of water in the surrounding guard cells. Increased water pressure makes the cells turgid (bloated). The guard cells maintain turgidity by accumulating ions (especially potassium) in them. Water then moves into them by **osmosis**. Guard cell walls are naturally thickened and the increasing pressure causes the elastic outer walls to stretch and the stomata to open. Sometimes under microscopic examination cellulose microfibres can be seen radiating around the circumference. These are called elliptical rings. Generally light causes the stomata to open and dark causes them to close. Unfortunately, these natural openings into the plant also enable bacterial and fungal spores to penetrate, which can cause disease, e.g. fireblight (*Erwinia amylovora*). Figure 8.1 depicts a stoma.

Aim

To investigate the variability of stomata between monocotyledon and dicotyledon leaves.

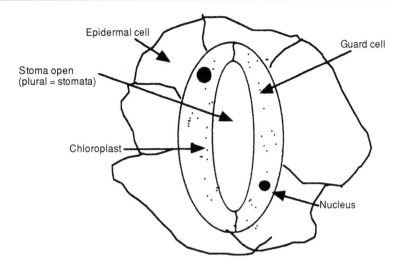

Figure 8.1
Parts of a stoma

Apparatus

nail varnish
slides
cover slips
lacto-phenol blue dye
dicotyledon leaves (e.g.
 Hedera or *Pelargonium*)

monocular microscope
monocotyledon leaves (e.g.
 Grasses, *tradescantia* or
 alstroemeria, Narcissus, most
 bulbs)
binocular microscope

Method

1 Observe the underside of a monocotyledon and dicotyledon leaf using a binocular microscope. You may be able to see the stomata at high magnification.
2 Paint two thin layers of nail varnish on the underside of each leaf, leaving to dry between coats. When dry, peel off the varnish film and place it on a microscope slide. Add a drop of dye and a cover slip (sometimes results are better without using the dye).
3 Observe the stomata slides under a monocular microscope.

Results

Draw labelled diagrams of your observations of monocotyledon and dicotyledon leaf stomata including epidermal cell, guard cell, stomata.

Conclusions

1 Estimate, based upon your observations, the number of stomata per square centimetre.
2 Are your specimens stomata open or closed?
3 How did the stomata vary between monocotyledon and dicotyledon specimens?
4 How many elliptical rings could be seen in each guard cell?
5 What is the function of the stomata?
6 For which plant process would the stomata functioning be important?

7 Are the stomata open during the day or at night?

8 How might the plant regulate the opening and closing of the stomata?

9 State the disadvantage of stomatal pores to plant health.

Exercise 8.4 *Transpiration*

Background

Transpiration is the process resulting from water uptake through the roots of the plant and passing, via the stem, to the leaves where it is lost by diffusion through the stomata into the air. Under hot shop lights or windy storage conditions plants may lose more water than they can take up. This results in wilting and the collapse of the leaves and stems as they become limp through the loss of water.

Transpiration is important to plants because it enables:

- water movement around the plant
- nutrients to move around the plant
- leaf cooling

The majority of water is moved as a result of transpiration. Ninety-five per cent of water loss is through transpiration. Transpiration facilitates the movement of water and nutrients. Nutrients are required at the sites of photosynthesis for which transpiration is essential.

The **transpiration stream** describes the movement of water through the plant and its loss through the leaves pulling further water up from the ground. A suction of 30–40 bars (1 bar = 1 atmospheric pressure) has been recorded in eucalyptus, due to transpiration from the leave surface pulling the water columns up through the xylem. The transpiration stream is the major force causing upward water movement. The column of water in the xylem vessels do not break under this strain, but rather act like a solid, so that the water transmits the pull throughout the whole of its length and the whole column moves upwards. It has been estimated that the suction force set up by the leaf cells as a result of the transpiration stream could raise water to a height of over 300 m.

Water loss through the epidermis of the leaf is called **evaporation**. About 5 per cent of water loss occurs directly from the epidermis cuticle by evaporation. Plants reduce this by producing a waxy coating on the epidermis. These are shiny and help to reflect radiation. Hairs also reduce water loss due to evaporation. Humid air gets trapped between the hairs lowering the humidity gradient between inside and outside the leaf, slowing down both evaporation and transpiration. Substances called antitranspirants may be used to spray onto leaves to reduce both transpiration and evaporation. Transpiration rates are influenced by a variety of environmental factors, including:

Solar radiation. On bright, sunny days transpiration increases and plants lose more water. This has implications for the amount of irrigation water required to be applied to the plant.

Air temperature. Ambient temperature is the surrounding background temperature. Transpiration increases with temperature. If we can therefore lower the temperature (e.g. shading or ventilation) we are able to reduce the water loss. For every 10°C rise the rate of transpiration rises by 50 per cent.

Air movements. Creating a diffusion gradient between dry and moist air allows diffusion from the leaf (e.g. drafts and central heating systems). Windbreaks greatly reduce water loss.

Humidity. Air can only take up certain amounts of water before it becomes saturated. As the air becomes wetter (humid) it can take up less water than dry air. If we therefore increase the humidity we can lower the plant's transpiration rate (e.g. misting, spraying or covering plants with a damp sheet). In understocked glasshouses the relative humidity (RH) may drop so low that plants may wilt even if well watered. The optimum relative humidity is 50–80 per cent. RH decreases by 10 per cent for every 1°C rise in temperature.

Soil/container moisture content. As soil moisture deficit increases, transpiration decreases, and the plant is under increasing stress and may commence to wilt.

Aim

To assess the transpiration rate of *Garrya elliptica*, under different environmental conditions.

Apparatus

Bubble potometer (see Figure 8.2) comprising:

bung and demi-john	reservoir of water
capillary tube	retort stand and clamps
Garrya elliptica cutting	ruler
hair drier	valve

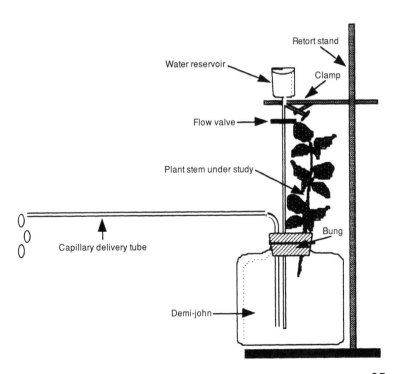

Figure 8.2
Bubble potometer

Method

1 Set up the apparatus as shown in Figure 8.2.
2 Open the reservoir valve and fill the demi-john until water backs down the capillary tube and drips roll off the end, then close the valve.
3 Mark the capillary tube 1 cm from the end.
4 Subject the plant to six different environmental conditions and record how much moisture is used over 5-minute time periods, by recording the distance travelled by the miniscus from the mark made in step 3.
5 It may be necessary to reset the potometer to the end (step 2) if transpiration rates are rapid and the miniscus advances towards the end of the capillary tube.

Results

Enter your results in the table provided. The actual volume of water travelling up the capillary tube may be calculated from the formula:

$$\pi r^2 h$$

Where π = 3.142, r = radius of capillary tube (1 mm), h = distance travelled by miniscus (mm)

Calculate the actual volume of water passing through the capillary tube over the 5-minute period. Show your working, following the format given in the results table.

Sample conditions	Transpiration loss of moisture over 5 minutes (mm)	Volume of water transpired (ml) $\pi = r^2 h$
Example environment	600 mm	3.14 × 1 × 600 = 1884 ml
Room temperature		
High hot air movement		
Outside in shade		
Outside in brightness		
Outside in a cold wind		
Other		

1 Produce a bar chart, on graph paper, to show how transpiration varies with environment.
2 Which environment produced the greatest rate of transpiration?

3 Comment on the significance of your answer to question 2 for garden centre retailed plants.

4 Substances called antitranspirants are available from garden centres. When sprayed over the plants, they produce a waterproof covering for leaves. Explain why antitranspirants are used in each of the following situations:

(a) preventing wilting in greenhouse crops
(b) assisting cuttings to root
(c) reducing the fall of needles from Christmas trees.

Conclusions

1 Define the term 'transpiration'.
2 Describe four environmental factors that affect the rate of transpiration.
3 Describe what happens to plant cells when plants suffer from a lack of water.
4 State two methods to relieve plants that are wilting.
5 Describe three possible ways in which water may rise from the ground roots to the top of a large tree and mention the relative importance of each mechanism.
6 Warm air currents produced by domestic heating systems have a drying effect on the air. Explain the effect this has on the transpiration rate of house plants.
7 Describe and explain how humidity can be controlled when raising plants.
8 Describe how the transpiration rate may affect the practices of the grower when:

(a) propagating from cuttings
(b) growing vegetables in an exposed area
(c) producing cucumbers under glass.

9 Explain how the rate of transpiration is affected by the following:

(a) mist propagation environment
(b) exposure to wind
(c) damping down
(d) reduction in temperature
(e) bright sunshine.

10 State six methods that you could use to reduce water loss from growing plants.
11 Describe how water moves:

(a) through the soil to plant roots
(b) across the plant roots to the xylem
(c) up the plants stem to the leaves
(d) out of the leaves into the atmosphere.

12 Describe, using a diagram, how you would carry out an experiment to test the rate of transpiration from a cutting.

Exercise 8.5 *The weather*

Backgound

In Britain the weather is a major force shaping the growth and appearance of plants. For example, the shape of growing trees is largely sculptured by wind action. Shading results in the loss of lower branches. They rarely exceed two hundred years of age and are mostly lost through storm damage. Similarly, a common feature of town houses with small fenced garden lawns, is algae and fungal disease growth. This is encouraged by the persistent wet conditions and high humidity created by high fences reducing air movement and slowing evaporation from the soil and transpiration from the grass.

Meteorological data is an extremely useful tool in maintaining the health and establishment of plants. For example, rainfall data is the basis of planning irrigation, and temperature data aids the timing of fertilizer applications. Weather forecasting is supported by information provided from a network of meteorological weather stations spread across Britain, reporting to the central Meteorological Office in Berkshire.

Many controlled environments used in plant production maximize plant response to the processes of photosynthesis, respiration and transpiration. This is through computerized control in the internal climate of the structure, in response to the daily weather conditions.

Aim

To gain an appreciation of the variety of techniques available to record British weather patterns.

Apparatus

A suitable weather station comprising:

anemometer   soil thermometers
grass thermometer   Stevenson's screen
rain gauge

Method

Visit the weather station and record the weather today.

Results

Enter your results in the table provided.

Recording	Value
Stevenson's screen: Maximum temperature	
Minimum temperature	
Soil temperatures: Grass minimum temperature	
Soil at 10 cm	
Soil at 30 cm	
Rain gauge: mm of rainfall	
Anemometer (wind speed)	

Conclusions

1 How does temperature vary between above and below ground?
2 What does this mean in terms of crop growth in spring?
3 What effect will today's weather have on the transpiration rate of plants?

9 Flower structure

Background

The blossom is a specialized shoot of a plant bearing the reproductive organs. Often brightly coloured to attract fertilizing insects, it is the primary organ involved with sexual reproduction. Once the stem apex has begun the shift from vegetative growth to flower development, the change is normally irreversible.

The flower is usually formed from four concentric rings (whorls), the sepals, petals, androecium and gynaecium, attached to a swollen base (receptacle) at the end of a flower stalk (pedicel). The female tissues are called gynaecium; the male tissue is called the androecium. Most other flower tissue is concerned with protection and attracting insects.

Figure 9.1 demonstrates a typical half-flower structure, showing the position of all the structures and their role in contributing to sexual reproduction.

Flower sexuality

Flowers contain either, male, female, or male and female sets of reproductive organs. These are classified as dioecious, monoecious or bisexual (hermaphrodite) respectively.

Dioecious

Separate male and female plants, i.e. having male and female flowers on separate plants. For fertilization purposes a male plant is usually set among a group of females; both are required if pollination is to take place as self-pollination is impossible. Not a

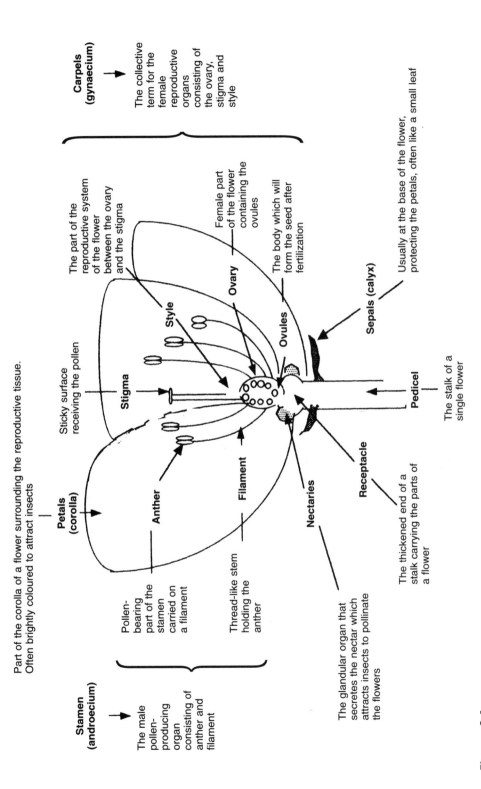

Figure 9.1
Typical flower structure

Carpels (gynaecium) → The collective term for the female reproductive organs consisting of the ovary, stigma and style

The part of the reproductive system of the flower between the ovary and the stigma

Female part of the flower containing the ovules

The body which will form the seed after fertilization

Style

Ovary

Ovules

Sepals (calyx) — Usually at the base of the flower, protecting the petals, often like a small leaf

Part of the corolla of a flower surrounding the reproductive tissue. Often brightly coloured to attract insects

Sticky surface receiving the pollen

Stigma

Petals (corolla)

Anther

Filament

Nectaries

Receptacle

Pedicel — The stalk of a single flower

Pollen-bearing part of the stamen carried on a filament

Thread-like stem holding the anther

The glandular organ that secretes the nectar which attracts insects to pollinate the flowers

The thickened end of a stalk carrying the parts of a flower

Stamen (androecium) → The male pollen-producing organ consisting of anther and filament

very common group. E.g. asparagus, date palm, Garrya, Ginkgo biloba, poplar (Populus), holly (Ilex), Pernettya, Skimmia japonica, willow (Salix), yew (Taxus), cucumber.

Monoecious

Having separate male and female flowers, but on the same individual plant. This is most common in wind-pollinated plants such as hazel were the catkins are the male flowers and the small buds are the female flowers. Self-pollination is unlikely. E.g. beech (Fagus), birch (Betula), hazel (Corylus), maize, marrow, oak (Quercus), sycamore (Acer pseudoplatanus), walnut, most conifers.

Bisexual or hermaphrodite

Flowers with both male and female organs on the same flower, so they do not require a separate pollinator unless the plant is self-sterile. E.g. Alstroemeria, buttercup, white deadnettle, bluebell, apple, pea and most plant species.

This chapter will investigate the following areas:

(a) function of the flower
(b) flower sexuality
(c) differences between insect- and wind-pollinated plants
(d) pollination and fertilization.

Exercise 9.1 *Flower dissection*

Aim

To gain competence at dissecting flowers and identifying their components tissues.

Apparatus

> fritillaria flowers
> hand lens
> dissecting kit

(suitable alternative flowers may be used, e.g. daffodil, lily and Alstroemeria)

Method

1 Place a fritillaria flower on a white tile and bisect the flower.
2 Draw a diagram of the half flower.
3 Dissect the flower and place each structural part on a piece of card identifying its name.

Results

Draw a labelled diagram of all the tissue present in the half-flower.

Conclusions

1 Define the terms

    (a) dioecious
    (b) monoecious
    (c) hermaphrodite.

2 Is the plant you have drawn dioecious, monoecious or hermaphrodite?
3 List an example plant that is:

    (a) dioecious
    (b) monoecious
    (c) hermaphrodite.

4 Draw a clearly labelled diagram of the 'typical' half-flower structure of a hermaphrodite (bisexual) flower with which you are familiar.

Exercise 9.2 *Structure and function*

Aim

To describe correctly the function of flower structure parts.

Apparatus

Previous illustraton of flower parts

List of structure parts:

anther	ovary	sepals
carpels	ovules	stamens
filaments	pedicel	stigma
nectar	petals	style
nectaries	receptacle	

Method

Match the correct flower tissue with the following description of its function.

Results

What is the:

    (a) thread-like stem that carries the anther in a stamen?
    (b) part of the stamen producing the pollen grains?
    (c) female tissue which when receptive is sticky and receives the pollen at pollination?
    (d) part of the reproductive system of the flower between the ovary and the stigma?

(e) female part of the flower which will develop into a fruit after fertilization?

(f) female reproductive organs consisting of the ovary, stigma and style?

(g) body that will develop into the seed after fertilization?

(h) part of the corolla of a flower surrounding the reproductive organs, often coloured to attract insects?

(i) male pollen-producing organ consisting of anther and filament?

(j) glandular organ that secretes the nectar?

(k) 'beverage of the gods' – the solution secreted by the glands that attracts insects to pollinate the flowers?

(l) tissue, often like a small leaf, at the base of the flower that protects the petals when in bud?

(m) stalk of a single flower?

(n) thickened end of a stalk carrying the parts of a flower?

Conclusions

1 Explain the difference between 'gynaecium' and 'androecium'.

2 Describe the typical arrangement of flower tissue.

3 Describe each of the following structures and state their role in the flower:

(a) stamens
(b) stigma, style and ovary
(c) ovules
(d) petals

Exercise 9.3 *Floral structures*

Background

Flowering plants belong to two main groups, the dicotyledons and monocotyledons. Examination of the flower parts will determine to which group they belong.

Dicotyledons: Flowers have distinct sepals, petals, stamens and carpels, in multiples of four or five.

Monocotyledons: Flowers do not have distinct sepals or petals. Instead there are one or two rings of petal like perianth segments. A perianth is a collective term used to describe the envelope of external floral parts, including the calyx and corolla. The combined petals and sepals of a tulip are called a perianth. The perianth segments of these flowers are in multiples of three.

Ovary – superior or inferior?

Superior: attached to the receptacle above the position of insertion of the corolla whorl and stamens.

Inferior: attached to the receptacle below the position of insertion of the corolla whorl and stamens, so that it can be seen below the flower in side view.

Aim

To compare and contrast the flower of a dicotyledon with that of a monocotyledon.

Apparatus

alstroemeria	lily
bluebell	poppy
buttercup	tulip
daffodil	dissection kit
daisy	hand lens

Method

1 Dissect and examine each flower.
2 Record your observations in the table provided.

Results

Enter your results in the table provided.

Floral features	alstroemeria	bluebell	buttercup	daffodil	daisy	lily	poppy	tulip
Petal segments joined or separate								
Number of petal segments								
Number of stamens								
Superior or inferior ovary								
Monocotyledon or dicotyledon								

Conclusion

Describe the floral differences between monocotyledonous and dicotyledonous plants.

Exercise 9.4 *Pollen investigation*

Background

Pollination is defined as the the transfer of pollen (the male sex cell), from the anthers to the female stigma. When ripe, the pollen sacs of the anther split open and expose the pollen which can then be dislodged. The pollen is either carried by insects, or blown by the wind, to the stigma of another flower. This leads to fertilization and fruit development resulting in new seed. The practice of dehiscing flowers (breaking open the anthers to release the pollen), is a widespread practice in horticulture. Similarly, the use of an electric bee, a vibrating probe shaking the flower to dislodge the pollen, encourages pollination in glasshouse crops. If pollination is necessary (e.g. in apples) it is important not to apply insecticides or fungicides at full flower.

Generally plants are adapted to enable either insect or wind pollination and this is reflected in the characteristics of the flower and the pollen grain produced. In the process of evolution, the structure and physiology of a flower has been modified in ways which improve the chances of successful pollination by insects or wind. These adaptations include the following.

Insect-pollinated plant adaptations

The sequence involves an insect visiting one flower, becoming dusted with pollen from the ripe stamen and visiting another flower where the pollen on its body adheres to the stigma.

General flower adaptations	Pollen characteristics
• brightly coloured petals to attract a variety of insects	• small anthers firmly attached to flower
• scent	• anthers lie within flower
• honey guides (dark lines) direct insects to nectar source	• anthers positioned for insect collision
	• large conspicuous flowers
	• smaller quantities of pollen produced than for wind pollination
• honey guides force contact with stamens and stigmas	• large pollen grain size
	• sticky pollen
• nectar	• spiked/barbed pollen enabling them to stick in clumps on insects

Wind-pollinated plant adaptations

Grasses, for example, are pollinated by air currents. First, the feathery stigmas protrude from the flower and the pollen grains, floating in the air, are trapped by them. Subsequently, the anthers hang outside the flowers, the pollen sac splits and the wind blows the pollen away. The sequence varies with species.

General flower adaptations	Pollen characteristics
• small inconspicuous flowers	• anthers large, loosely attached to filament
• petals small – plain, often green	• stamens hang outside flowers, exposed to the wind
• no scent	
• no nectar	• anthers and stigma hang outside the flowers so the slightest air movement shakes them
• whole flower loosely attached	
• feathery stigmas act as a net which traps passing pollen grains	• large quantities of smooth, lightweight pollen are needed because it is less reliable
	• smooth, aerodynamic grains

Aim

To investgate the differences between wind- and insect-carried pollen.

Apparatus

    pollen from grass (or other wind-pollinated plants, e.g. pine)
    pollen from fritillaria (or other insect-pollinated plants,
        e.g. lily, iris, daffodil)
    binocular microscope

Method

Observe the pollen under the microscope and compare and contrast their structures.

Results

Describe the differences in the two types of pollen. Draw a diagram of your observations.

Conclusions

1 State which flower's pollen grain was the largest in size and explain the significance of this.
2 Describe and comment on the difference in shape of the two types of pollen.
3 Explain why the fritillaria pollen was sticky and grouped in clumps.
4 Describe four adaptations which exist in wind-pollinated flowers.
5 State what is meant by the term 'pollination' and explain its importance to flowering plants.
6 Name two example flowers that are pollinated by wind.
7 List three ways by which some plants are adapted to prevent self-pollination.
8 List four structures which exist in flowers and describe their function, if any, in insect-pollinated plants.

Exercise 9.5 *Insect-pollinated plants*

Background

Bees pollinate flowers allowing fertilization to take place. They feed on the nectar and collect pollen from the stamens. Normally a group of insects will work a group of flowers methodically to collect as much pollen as possible.

Aim

To observe the variety and frequency of insect pollination of flowers.

Apparatus

    graph paper
    stop watch

Method

1 Select a small patch of named flowers (e.g. heathers) for observation.
2 Observe all the flowers in the patch over 15 minutes, and record:

(a) the number of bees visiting the patch
(b) the time spent by each bee at the patch
(c) the number of flowers visited by the bee during this time.

Results

Enter your results in the table provided.

Observations	Patch 1	Patch 2	Patch 3
Number of bees visiting patch			
Average time spent by each bee at the patch			
Average number of flowers visited by each bee			

Produce a bar graph summarizing your data.

Conclusions

1 What was the average result obtained?
2 How might a strong wind have affected your results?
3 What other insects were involved?
4 Name a flower that is pollinated by a large insect.
5 Describe four adaptations that exist in insect-pollinated flowers.

Exercise 9.6 *Fertilization*

Background

The fusion of the male pollen nucleus with the female ovule nucleus to form a seed is called fertilization.

Fertilization in plants follows pollination. The pollen grain absorbs sugar secreted by the stigma. A pollen tube is formed which grows down through the style. On reaching the ovary, the tube grows to one of the ovules and enters through a hole, called the micropyle. The tip of the pollen tube breaks open in the ovule, and the nucleus, which passes down the tube, enters the ovule and fuses with the female nucleus. Each egg cell in the ovule can be fertilized by a male nucleus from a separate pollen grain. A fertilized ovary is called a **fruit**.

Sometimes fertilization can occur without pollination. This creates parthenocarpic fruit. For example, seedless grapes and bananas are fruit set without pollination occurring. They are chemically treated with a hormone to induce fruit set. Flowers that suffer frost damage (e.g. pear crops) will have their styles damaged, which will prevent the growth of the pollen tube and on inspection the style will be black; fertilization cannot take place. But when sprayed with Berelex (gibberellin hormone) they form parthenocarpic fruit. The fruit will be misshapen, but flavour is not affected.

After fertilization, most of the flower parts (petals, stamen, style and stigma) wither away. Sometimes traces of them may remain (e.g. the receptacle swells and becomes succulent as in apples).

In all plants the main change after fertilization involves the ovary and its contents. Generally, the ovary appears greatly swollen. Inside the ovule, cell division and growth produces a seed containing a potential plant or embryo, and the outer ovule wall forms the seed coat (testa). Water is withdrawn from the seed making them dry and hard, enabling them to withstand unfavourable conditions. If conditions are suitable the seed will germinate; if not they will remain dormant until the conditions are right.

Aim

To investigate fertilized and unfertilized flowers.

Apparatus

daffodils A and B (or any two flowers)
microscope
dissection kit

Method

1 Examine the two daffodils.
2 Dissect both ovaries.

Results

1 Describe the differences between the two daffodils.
2 Draw a diagram of each ovary showing how they differ.

Conclusions

1 Explain which plant has been fertilized giving reasons for your answer.
2 Define the term 'fertilization'.
3 Explain how 'pollination' differs from 'fertilization'.
4 Describe the changes that occur in flowers after fertilization.
5 Describe the methods by which poor fertilization may be improved in horticulture.
6 Briefly describe the sequence of events leading to fertilization and fruit and seed formation.

Part Two Soil science

10 Soil texture

Background

Soil texture is the single most important soil property. Most aspects of soil management are related to it. Texture is the relative proportions by weight of sand, silt and clay in the soil. This is a basic inherited property of a soil that cannot be changed easily by cultivation.

The mineral component of soils can be split into three types (sand, silt, and clay) defined solely by size. There are several different classifications but the one now being used by ADAS (Agricultural Development and Advisory Service), a division of the Ministry of Agriculture Fisheries and Food (MAFF), is as in Figure 10.1 and Table 10.1.

Technically, anything larger than 2 mm becomes a stone and is not 'soil'. In addition if a soil contains more than 45 per cent

Particle	Symbol	Subdivision and textural feel	Particle size (diameter) mm
Sand	S	**Coarse:** Easily seen with the naked eye. Feels and sounds very gritty	0.6–2.0
		Medium	0.2–0.6
		Fine: Only slightly gritty. Difficult to detect in soils and easily missed	0.06–0.2
Silt	Z	Silky, smooth often buttery when wet	0.002–0.06
Clay	C	Stiff to the fingers; very sticky when wet	0–0.002

Table 10.1
Assessing particle size

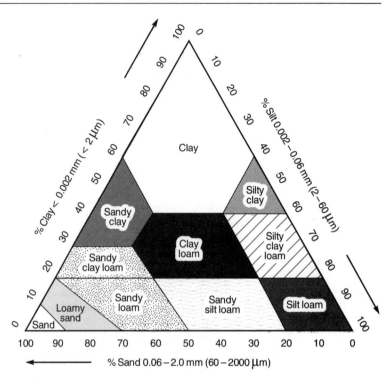

Figure 10.1
Proportions of sand, silt and clay in soil

organic (peaty) matter it is classified as an organic soil rather than a mineral soil and different methods to determine texture must be employed.

The symbol 'L' may be used to represent a loamy soil (e.g. ZCL – silty clay loam). Soil texture may be determined by working the soil between the fingers. Each textual class can be distinguished by its moulding properties. There are a range of sizes present within these classes and texture assessment helps us to understand the working properties of our soil including:

(a) Is the soil easy to dig? (structure)
(b) Will it grow plants easily? (cultivations)
(c) Will it hold water easily?
(d) Will it be 'fast' or 'late' to warm up in the spring?
(e) Will it drain well?
(f) Application rates of pesticides and herbicides.

The exercises in this chapter are designed to introduce the technique of hand texture assessment using the ADAS system modified in conjunction with the Soil Survey of England and Wales in 1986. This is the method used extensively in land management disciplines in the UK.

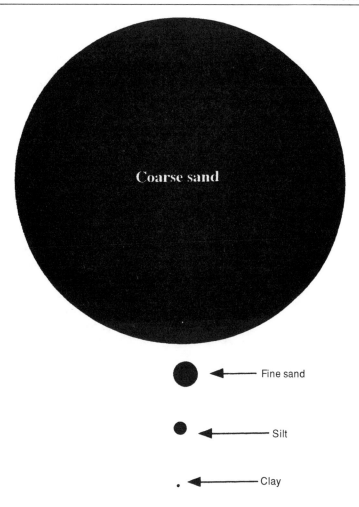

Fine sand

Silt

Clay

Figure 10.2
Particle size

Exercise 10.1 *Properties of soil particles*

Background

Soil texture is the relative proportions of sand, silt and clay in the soil. The particles of these elements vary in size (see Figure 10.2).

Sandy soils (0.06–2.0 mm)

Feel gritty. These soils are easy to dig but it is often difficult to grow 'good' plants in them. They are well drained and aerated. These soils do not hold plant nutrients at all well. They have a low water-holding capacity and water drains through the soil quickly.

Sandy soils are prone to drought and in dry weather they dry out quickly. Plants find it difficult to get enough water. It is important to remember that sand particles cover a wide range of diameters. Coarse sand will drain much faster than medium or fine sand. This is important when choosing sands for improved drainage or for formulating loamless compost mixes.

Silty soil (0.002–0.06 mm)

These are stone-free, deep soils. They feel silky and soapy. Not very sticky and not very gritty. They can be difficult to cultivate and may cause drains to block as they 'cake' the porous drains with layers of sediment.

Their workability is often improved by adding bulky organic matter, compost or farm yard manure. Normally they hold enough water and nutrients to sustain plant growth.

Clay soils (less than 0.002 mm)

These feel sticky and buttery when wet. They are very difficult to dig when dry. The best time to dig them is when they are moist, but not too wet. Clay particles have good colloidal (chemical) properties and are able to hold onto large quantities of nutrients making them very useful for growing good plants.

They are often called '**cold**' or '**late**' soils because the water which lies in the spaces between the soil particles (pores) warms up slowly in spring. As they absorb water they become very sticky and unworkable. When they dry out, they shrink, becoming hard and cracked.

Loams

These soils provide a mixture of soil particles and are considered by many people as the best soils to grow plants in. They contain roughly equal proportions of sand, silt and clay.

Aim

To distinguish by feel the relative properties of sand, silt and clay.

Apparatus

| sandy soil | clay soil |
| silty soil | water bottle |

Method

1 Place a spoonful of soil in the palm of your hand.
2 Rub the soil together between finger and thumb.
3 Record what the soil feels like.
4 Think of a similar material which feels the same as the soil. This helps you to remember the characteristic.

Results

Record your results in the table provided. Suggested comparable materials include salt, pepper, talcum powder, plasticine, soap, butter and others.

Soil type	Texture (feel) characteristics	Material with a similar feel
Sand		
Silt		
Clay		

Conclusions

1 State the accepted size range for the classification of soil particles into:

(a) coarse sand
(b) medium sand
(c) fine sand
(d) silt
(e) clay.

2 Describe the properties of each of the following:

(a) sand
(b) silt
(c) clay.

3 Select the appropriate word to describe how clay particles can be recognized when handled between the fingers and thumb:

(a) crumbly
(b) gritty
(c) silky
(d) sticky.

4 State two benefits of clay as an ingredient in soil.
5 Describe how sand may be identified by feel and state one advantage and one disadvantage of it as a constituent in soil.

Exercise 10.2 *Features of coarse, medium and fine grade sand*

Background

Sandy soils are mostly quartz and silica mineral grains resistant to weathering. They contain large air spaces and have poor water-holding capacities. For this reason medium sand, particularly, is useful as a top dressing on a newly spiked lawn to increase aeration. Because of the large diameter size of coarse sand, it may sometimes be useful to mix with a clay soil and increase the number of small pores aiding drainage and aeration. They are chemically inactive (inert) making them sterile. Because of the lack of excess moisture and large air spaces, they warm up quickly in the spring but may be prone to drought in summer.

Sands fall into three size ranges:

Coarse: 0.6–2.0 mm (600–2,000 μm)
Medium: 0.2–0.6 mm (200–600 μm)
Fine: 0.06–0.2 mm (60–200 μm)

Well-drained sandy soils produce the ideal conditions for the breakdown of organic matter by bacteria, and are often referred to as '**hungry**' soils because they require a great deal of supplementary nutrient feeding (fertilizers) and mulching.

Aim

To distinguish between coarse, medium and fine grade sand.

Apparatus

binocular microscope
glue
coarse, medium and fine grades of sieved sand

Method

1 Rub each sand grade together between finger and thumb.
2 Record what the sand grade sounds and feels like.
3 Think of a similar material which feels the same as the grade. This may help you to remember the characteristics.
4 Observe and describe each grade under the binocular microscope.
5 Glue a small sample onto the sheet for future reference.

Results

Record your results in the table provided.

Sand grade	Textural (feel) character-istics	Audible (sound) character-istics	Microscope observations	Sample specimen
Coarse				
Medium				
Fine				

Conclusions

1 Which sand would you use as a top dressing on an ornamental lawn?
2 Which sand would you use to improve the properties of a clay soil?
3 What is the effect of adding organic matter on sandy soils?
4 Which of the following is essential to allow water to drain through soil:

(a) a continuous film of water around soil particles
(b) a large number of small pores
(c) a sufficient number of large pores
(d) plenty of stones?

Exercise 10.3 *Sand grade investigation*

Background

Often it is useful to use sand as a surface material either on which to stand container-grown plants, or as a root zone mixture for sports turf and lawn top soils. In all these situations a special blend of different sand grades should be used. The resulting 'mixture' should have enough large pore spaces to allow water to drain away, but enough small pores to transport water by capillary action (siphoning) up to the plant roots. Once such material is called 'Efford' sand, designed by Efford environmental research station as a standing material for container-grown plants. Other materials that you may like to investigate include the Texas University specification for the United States Golf Association (USGA) golf green root zone mixture.

Aim

To observe the substantial variations in sand grades present in Efford sand.

Apparatus

> Efford sand
> sieve nest
> measuring cylinders (10, 100 and 500 ml)
> 500 ml beaker

Method

1 Measure out approximately 500 ml of sand.
2 Arrange sieve nest in the following order: receiver tray, 0.06 mm (fine), 0.2 mm (medium), 0.6 mm (coarse), 2 mm (stones), lid.
3 Place 500 ml of sand onto the top sieve (2 mm) and shake for 5 minutes.
4 Separate sieve nest.
5 Place each grade of sand into a measuring cylinder and record its volume.
6 Calculate the percentage of each grade within your 500 ml sample volume.
7 Save some of each grade sand in a weigh boat for use in Exercise 10.2.

Results

Enter your results in the table provided.

Particle sizes	Volume within 500 ml sample (A)	% of sand grade within 500 ml sample = A/500 ml × 100
Coarse sand 0.6–2 mm		
Medium sand 0.2–0.6 mm		
Fine sand 0.06–0.2 mm		
Receiver tray < 0.06 mm		

Conclusions

1 What grade of soil falls through all the sieves onto the receiver tray?
2 What properties does Efford sand possess that are useful in horticulture?
3 Which of these statements explains how in spring a heavy soil warms up more slowly than a sandy soil:

  (a) it becomes colder during the winter period
  (b) it has a lower water-holding capacity
  (c) surplus moisture does not drain away so quickly
  (d) warm air cannot enter so easily?

Exercise 10.4 *Soil texture assessment*

Background

The mineral fraction of the soil is the collective term for sand, silt and clay particles. The relative amounts of these particles is defined as soil texture. The ability accurately to determine the soil texture is a skill that horticultural-ists find particularly challenging but one that will pay great dividends when properly mastered. Many aspects of land management rest upon its determination in the field situation. For example, pesticide and herbicide application rates, nitrogen fertilizer recommendations and general skills of manipulating the growth medium to enhance plant growth and develop-ment rest upon the foundation of soil texture.

The ability to determine soil texture by hand assessment is a skill that will require regular practice. It is a good idea to build up a collection of soils with different textures and repeatedly practise on these samples. The purchase of a local soils map from the Land Research Centre (previously Soil Survey of England and Wales) will be a good starting point. A more accurate determination of soil texture may be obtained by submitting a soil sample to a laboratory for analysis. However, the hand texture assessment method will be adequate in most cases. It is suitable for most soils, except those with above 45 per cent organic matter which are termed 'organic' soils rather than mineral soils.

Aim

To develop skills to correctly conduct a hand texture assessment on different soil types.

Apparatus

  Figure 10.1
  Figure 10.3
  soils of different textures labelled 1–10
  water bottle

Method

Using Figure 10.3 assess the textures of the soils present in the sample buckets.

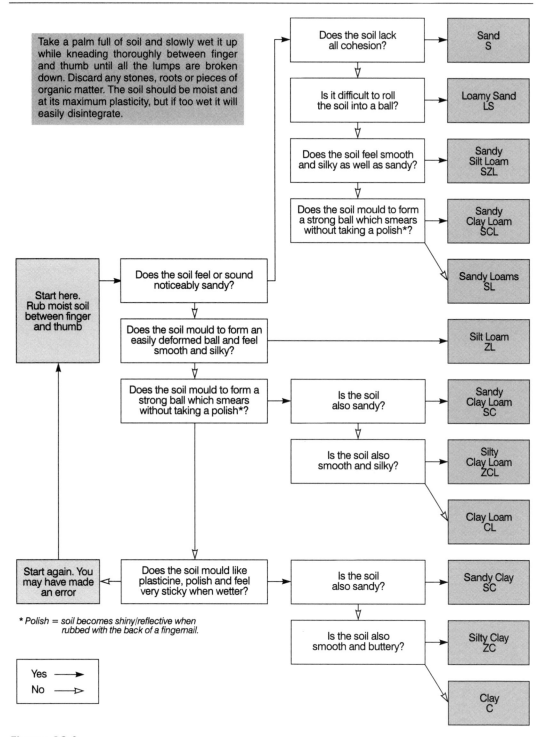

Take a palm full of soil and slowly wet it up while kneading thoroughly between finger and thumb until all the lumps are broken down. Discard any stones, roots or pieces of organic matter. The soil should be moist and at its maximum plasticity, but if too wet it will easily disintegrate.

Does the soil lack all cohesion? → Sand S

Is it difficult to roll the soil into a ball? → Loamy Sand LS

Does the soil feel smooth and silky as well as sandy? → Sandy Silt Loam SZL

Does the soil mould to form a strong ball which smears without taking a polish*? → Sandy Clay Loam SCL

Sandy Loams SL

Start here. Rub moist soil between finger and thumb

Does the soil feel or sound noticeably sandy?

Does the soil mould to form an easily deformed ball and feel smooth and silky? → Silt Loam ZL

Does the soil mould to form a strong ball which smears without taking a polish*? → Is the soil also sandy? → Sandy Clay Loam SC

Is the soil also smooth and silky? → Silty Clay Loam ZCL

Clay Loam CL

Start again. You may have made an error

Does the soil mould like plasticine, polish and feel very sticky when wetter? → Is the soil also sandy? → Sandy Clay SC

Is the soil also smooth and buttery? → Silty Clay ZC

Clay C

* Polish = soil becomes shiny/reflective when rubbed with the back of a fingernail.

Yes ⟶
No ⇨

Figure 10.3
Field guide to soil texture assessment by hand

Results

Enter your results in the table provided.

Soil sample number	Texture assessment determination
1	
2	
3	
4	
5	
6	
7	
8	
9	
10	

Conclusions

1 How many textural classes are there?
2 Define the term 'soil texture'.
3 Explain the relationship between soil texture and temperature in spring.

Exercise 10.5 *Soil texture assessment of personal plot sites*

Background

Now that some practice has been gained at the principles of determining soil texture it will be necessary to apply these techniques to a field situation. This may be to investigate the soil from a garden, demonstration plot, allotment, sports field, cricket wicket, golf green or other area in which you have a personal interest.

Aim

To determine the texture of site top and subsoils using the hand texture assessment method and link your findings to the subsequent management practices needed.

Apparatus

Figure 10.3
plot's top soil
plot's sub soil

Method

Using Figure 10.3 assess the textures of the soils in your chosen site.

Results

Enter your results in the table provided.

Plot sample	Texture assessment
Top soil	
Subsoil	

Conclusions

Comment on how the soil texture of your site will effect the management of it in the following areas:

(a) water-holding capacity
(b) aeration
(c) temperature at different times of year
(d) nutrient availability and need for fertilizer
(e) need for organic matter or mulches
(f) drainage
(g) workability.

11 Soil structure and profile assessment

Background

The study of soil in the field

A horticulturist may think of soil as only the top 30 cm of material under cultivation. A civil engineer will think of soil as several metres of unconsolidated material overlying solid bedrock. However, a soil scientist will classify soil as the material formed by the action of some organic agency on the original rock, or parent material. In Britain this usually involves the study of the top 1–2 m of soil overlying unaltered parent material. In the tropics, by comparison, ten times this depth may be required to include the whole soil.

The cross section of the soil, seen on the face of a pit, quarry or other cutting, is called the soil profile. It consists of layers of material (soil horizons), each of which appears to have different properties from the horizons above and below. These horizons are the product of soil formation processes or 'pedogenesis', and should be distinguished from the layering simply inherited from the stratification of the parent material.

A soil profile is a vertical slice through the soil, usually obtained by digging an examination pit 1 m deep. It may be divided simply into top soil and subsoil as in Figure 11.1.

The study of the properties of these horizons may tell us a considerable amount about the characteristics of the particular soil and its agricultural or horticultural potential. For a detailed assessment of properties, including structure, a small hole should be dug to reveal the soil profile. The soil is then examined to a depth of 1 m.

The exercises in this chapter are designed to introduce you to soil profile structures and to enable you to interpret your observations to make recommendations on how the soil should

Litter layer	Characteristics of a good management
Top soil 0–15 cm	• crumb structure • high organic matter content • highly fertile • lots of root development • lots of soil organisms • freely draining
Sub soil 15–30 cm	• coarse structure • low organism populations • low organic matter content • freely draining
Parent material 30+ cm	• underlying rock geology structure

Figure 11.1
Simple soil profile

be improved/managed to achieve good plant growth. The following areas will be investigated:

- soil structure
- general background to site under investigation
- procedure for digging a soil inspection pit
- examination of soil profile pit
- production of a record sheet
- conclusions and horticultural significance of the findings.

Soil structure

The soil structure refers to the binding together of soil particles into natural (peds) or artificial (clods) structural units of different shapes and sizes.

There are seven different structural units that can be easily recognized in the field. These are illustrated in Figure 11.2.

Structure units may be further divided into very coarse, coarse, medium and fine size units. An easy way to become familiar with structure units is to go out into the garden and collect a small handful of soil from the soil surface. Compare the shapes with those in Figure 11.2. Another technique is to dig a spade full of soil and turn it over in the air before allowing the soil to fall off the spade and hit the ground. You should find that the soil will fracture along its natural fissure lines as it breaks up into its structural units on impact.

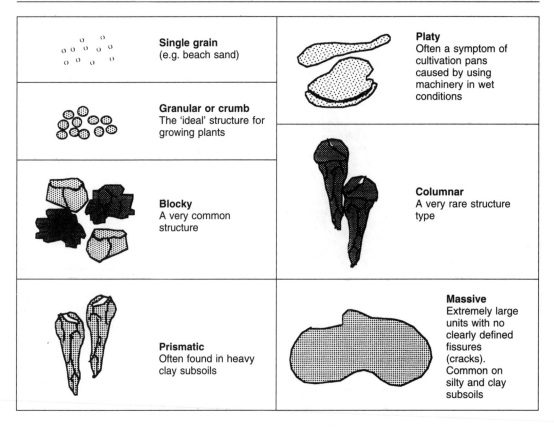

Figure 11.2
Soil structure types (not to scale)

Exercise 11.1 *Soil structure and profile assessment*

Aim

To act as an introduction to soil profile structures; to interpret observations from the assessment; and to make recommendations on how the soil should be improved/managed to achieve good plant growth.

Apparatus

metre rule	clip board	paper
pencils	spade	plastic bags
sample bags	labels	forks
augers	trowel	wash bottles
buckets	BDH pH kits	

Method

Select a site for investigation. A description of a soil profile involves the following steps.

Stage I   General background

Collect the following information:

1 a description of the site with map reference if available
2 details of vegetation
3 topography and drainage (e.g. shape of the land, undulating, rolling, flat, sloped, etc.)
4 local geology
5 signs of erosion
6 climate data (e.g. rainfall).

Stage II   Production of profile inspection pit

The procedure when digging and filling a soil profile pit is as follows:

7 Choose a position away from gateways and tracks.
8 Dig a pit approximately 1 m × 1 m × 1 m deep.
9 When digging the inspection pit care should be taken to:

(a) cut the turf so that it can be replaced easily and neatly
(b) keep the top soil and subsoil separate so that each can be returned to its own position without mixing them
(c) **clearly mark** where the hole is – use a long stick or cane with a cloth or flag, or rope the area off
(d) regularly consolidate the soil when refilling the pit
(e) replace all the soil in the hole, leaving the site 'proud' and allowing for settlement.

10 While digging:

(a) note hardness and compactness of the different horizons
(b) check for stones or pans if spade hits resistance.

11 Do not leave unmarked pits in fields. If, for any reason, you wish to leave a pit unattended you must inform the land owner and obtain their permission.
12 **Never** leave pits unattended in fields containing livestock.
13 When opening a profile pit in an orchard, remove the top 2–3 cm of soil (the herbicide layer) and keep it separate. It should be returned to the top of the soil after the profile pit has been filled in.

Stage III   Examination of soil profile pit

14 Clean a face of the pit for inspection.
15 Note the number of different horizons in the soil, the size of each horizon and the depth at which they occur.
16 Pattern of root development and penetration – note signs of horizontal rooting, deformed roots, frequency of fine roots, presence of roots within or between structure units, etc.
17 Note depth and frequency of worm channels.
18 Note points where water seeps into the pit.
19 Take samples of top soil and subsoil for chemical analysis.
20 For each horizon estimate the following characteristics, and record on your result sheet:

Horizon designation (top soil/subsoil)
Horizon depth and thickness

Colour
Texture   (use Figure 10.3)
Structure   (blocky, crumb, platy, prismatic, single grain, massive)
Pans
Concretions   (manganese, calcium carbonate, etc.)
Mottling   (orange rust colours, indicating seasonal waterlogging)
Gleying   (bluey-grey colour, indicating permanent waterlogging)
Fissures   These are natural cracks in the profile.

Fissures	**Porosity** (implied from fissure size)
Very fine <1 mm wide	(very slightly porous)
Fine 1–3 mm	(slightly porous)
Medium 3–5 mm	(moderately porous)
Coarse 5–10 mm	(very porous)
Very coarse >10 mm	(extremely porous)

Stones

Abundance	**Size**
Stoneless (<1%)	Very small stone (2–6 mm)
Very slightly stony (1–5%)	Small stones (6 mm–2 cm)
Slightly stony (6–15%)	Medium stones (2–6 cm)
Moderately stony (16–35%)	Large stones (6–20 cm)
Very stony (36–70%)	Very large stones (20–60 cm)
Extremely stony (>70%)	Boulders (>60 cm)

Organic matter (roots and animals)

pH   Use a BDH pH testing kit or similar item (see 'Measuring
                 soil pH' exercises)

Any other relevant information (e.g. a reclaimed rubbish dump,
previously used for cattle, small garden subject to excessive dog urea,
etc.).

Results

Enter your results in the blank 'Soil Profile Examination Record Sheet' (Figure
11.3). A completed example result sheet is included for reference (see
Figure 11.4).

Conclusions

A report containing observations, deductions and management decisions
should be prepared. A diagram of the profile should be drawn and
interpreted for management in terms of:

- ease of cultivation
- soil water-holding capacity
- drainage
- aeration works and subsoiling
- stability of top soil.

The report should also contain suggestions for possible planting for the site
examined along with reasons for this choice. Suggestions to improve the
site for planting may also be included. Before finally deciding on the plants
to be grown, the nutrient status of the soil should be considered and a
fertilizer recommendation made.

Soil Structure and Profile Results Sheet

Name: _____   Date: ____ / ____ / ____

Location: _____   Crop: _____

_____   _____

Depth cm	Sketch	Profile description
0		
5		
10		
15		
20		
25		
30		
35		
40		
45		
50		
55		
60		
65		
70		
75		
80		
85		
90		
95		
100		

Figure 11.3
Soil profile examination record sheet

Name: Pedologic Landscaping   **Date:** 04.11.97

Location: Plot 1, Sandy Lane project   **Crop:** Mixed trees and shrubs
Gardner's City
Bloomingtown

Depth Sketch   **Profile description**
cm   Surface – Light brown silt loam, Angular blocky. Very small stones 6–15%, slight capping pH 6.0–6.5

Colour:	Brown
Texture:	Sandy clay loam
Structure:	Course blocky, friable
Porosity:	Slightly
Fissures:	Few
Organic matter:	Low with black rotting material
Stones:	Small (2–6 mm) 10–15% abundance
Fauna:	None visible
pH:	6.0

Top soil

– 34 –

Colour:	Orangey brown	Organic matter:	Some black rotting o.m. present
Texture:	Silty loam	Stones:	None
Structure:	Coarse platy	Fauna:	None visible
Porosity:	Very slight	pH:	6.0
Fissures:	Very few		

Subsoil showing classic signs of cultivation pan

Colour:	Orangey brown	Porosity:	Very porous	Stones:	Medium (2–6 cm)
Texture:	Sandy silt loam	Fissures:	Many		70% abundance
Structure:	Structureless sand	Organic matter:	None	Fauna:	None visible
	and gravel vein			ph:	6.5

67 –

Colour:	Reddy orange
Texture:	Silty clay loam. Severe mottling from 83 cm
Structure:	Structureless massive. Very compacted
Porosity:	None
Fissures:	None
Organic matter:	Some black rotting material
Stones:	None
Fauna:	None visible
pH:	6.5

Figure 11.4

Example of soil profile examination record sheet

Interpret the significance of your findings to horticulture by completing the comments below.

1 What are the problems with this soil?
2 What are the benefits from this soil?
3 What is the management practice required to improve this soil?

    (a) cultivation
    (b) soil water-holding capacity
    (c) drainage
    (d) aeration works and subsoiling
    (e) stability of top soil
    (f) other.

4 Name three suitable plants/crops for use on this site.
5 Why are the top soils and subsoils kept separately?
6 Explain what mottling and gleying tell us about the soil.
7 Define the term soil 'structure'.
8 Distinguish between soil texture and soil structure.

12 Soil water

Background

The soil is composed of solid particles, air spaces and water, all of which are essential for healthy plant growth. However, different soil texture groups will have varying proportions of these properties, which may influence the growth rate of plants. Variations in particles sizes, for instance, greatly modifies the number and size of pore spaces in the soil and will subsequently influence the ability of the soil to drain freely. These exercises will investigate aspects of the water content of a variety of soils and also the consequential response of plants to over and under watering.

Soil water can exist in different forms in a soil profile. Some simple definitions may be useful.

Saturation point: All soil pores are full with water (water-logged).

Field capacity: Large pores are drained of water but small pores around particles contains films of water. This is the maximum amount of water that a soil can hold after the excess has drained away under the force of gravity.

Drainage water: The volume of water between saturation point and field capacity, which drains under the force of gravity (also called gravitational water).

Available water: Water in the soil which plants can take up, held between field capacity and permanent wilting point.

Permanent wilting point: The water content of a dry soil at which the water is held by the soil particles too tightly to be released by plant roots. Plants are therefore unable to draw water. Irreversible wilting follows and the plant will die.

Exercise 12.1 *Available water-holding capacity*

Background

The available water-holding capacity (AWHC) is a measure of soil available water content in grams per cubic centimetre ($g\,cm^{-3}$), and is often expressed as a percentage. Water is held on the surface of soil particles. Therefore the soil with the largest surface area will have the greatest available water-holding capacity. In practice this means that clay soils hold more water than sandy soils and will require less irrigation or watering in the hot summer weather. For this reason clay is often used as an ingredient in container compost, particularly where the interval between watering is likely to be great. However, they may require help to drain excessive winter rainfall away.

Aim

To determine the available water-holding capacity, at field capacity of sand and clay soils (approximate method).

Apparatus

analytical balance	pen knife
clay soil	polystyrene beaker (2)
filter paper	200 ml beakers (2)
marking pen	reservoir (washing up bowl)
measuring cylinder	sandy soil

Method

You may find it helpful to draw a labelled diagram of the apparatus. Record all readings in the relevant place in the results table (letters in brackets refer to rows in the table).

1 Using the balance determine the weight (a) of an empty polystyrene beaker.
2 Mark two polystyrene beakers with your initials and soil type (sand or clay)
3 Calculate the volume (b) of the polystyrene beaker using the following procedure:

    (i) Fill beaker with water.
    (ii) Pour contents into measuring cylinder.
    (iii) Record volume of beaker (b) in ml.

4 Fill one cup with dry sandy soil, firming regularly.
5 Fill second cup with dry clay soil, firming regularly.
6 Weigh each filled cup on the analytical balance. Record the result in row (c).
7 Calculate the weight of soil in the beaker, recording the result in row (d) ((c) – (a)).
8 Pierce bottom of each beaker with a knife to create drainage holes.
9 Bring each soil to saturation point by slowly lowering beakers into the water reservoir allowing the soil to become wetted from below. All the pores should now be filled with water.

10 When saturated (i.e. water floats on surface of soil and all pore spaces are full), gently place a piece of folded filter paper against the bottom of each beaker to prevent water draining away, and quickly transfer beaker into a 200 ml collecting beaker, and allow to drain freely.

11 When drainage stops record the volume of drainage (gravitational) water held in the collecting beaker (e). Your soil is now at field capacity.

12 Weigh beaker and soil at field capacity (f).

13 Determine the weight of soil at field capacity (g) ((f) – (a)).

14 Determine the weight of available water content held at field capacity (h) ((g) – (d)).

15 Calculate the available water-holding capacity (i), by dividing the weight of available water content (h) by the sample volume (b), noting that the density of water is $1.0\,g\,cm^{-3}$ (cm^3 = 1 ml., i.e. 1 g of water ≈ 1 ml volume).

16 Express your finding as a percentage of available water-holding content (j) using the formulae:

$$\text{\% available water-holding capacity} = \frac{\text{available water content weight (h)}}{\text{volume of soil contained in beaker (b)}} \times 100$$

Results

Enter your results in the table provided.

	Measurements	Example soil	Soil A	Soil B
(a)	Weight of empty polystyrene beaker (g)	20 g		
(b)	Volume of polystyrene beaker (ml)	250 ml		
(c)	Weight of beaker + soil (g)	265 g		
(d)	Weight of soil (c) – (a) (g)	265 – 20 = 245 g		
(e)	Volume of drainage water (ml)	50 ml		
(f)	Weight of soil + beaker at field capacity (g)	315 g		
(g)	Weight of soil at field capacity (f) – (a) (g)	315 – 20 = 295 g		
(h)	Available water content (g) – (d) (g)	295 – 245 = 50 g		
(i)	Available water-holding capacity (h/b) $(g\,cm^{-3})$	$\frac{50}{250}$ = $0.2\,g\,cm^{-3}$		
(j)	% available water-holding capacity $\frac{(h) \times 100}{(b)}$	$\frac{50 \times 100}{250}$ = 20 per cent		

Conclusions

1 Which soil type had the largest available water-holding capacity?
2 Explain your answer to Question 1.
3 How might the results have varied for a silt loam textured soil?
4 Which soil became wetted up quicker? And why?
5 Which soil will require more frequent watering?
6 Explain which soil is more likely to be drought resistant in the summer months.
7 How will the variations in available water-holding capacity influence the establishment or regrowth of plants in the spring months?
8 Describe how the suitability of these soils would influence the choice of plants to be grown?

Exercise 12.2 *Effects of waterlogged soils*

Background

There are many reasons why a plant may come to be growing in waterlogged soil. It may be due to compaction panning preventing free drainage, or as a result of overwatering. The effects on plant growth (physiology) can be dramatic. Plants may appear stunted, roots may die off and in the absence of oxygen, anaerobic (without oxygen) respiration may occur producing toxic alcohol as a waste product (see Chapter 6). This may be accompanied by the production of hydrogen sulphide gas producing a rotten egg smell. There are also many diseases carried by water including botrytis and damping-off.

Aim

To demonstrate the negative effects on plant growth of waterlogged soils.

Apparatus

clay loam soil
oil seed rape seed
seed trays (2)
water

Method

You may find it helpful to draw a labelled diagram of the plant appearances.

1 Prepare both seed trays with soil and sow with suitable seeds.
2 Water tray A, normally.
3 Overwater Tray B .
4 After 7 days record your results.

Results

Enter your results in the table provided.

Treatment	No. of seeds sown (X)	No. of successful germinating seeds (Z)	% germination $= \dfrac{Z}{X} \times 100$	Physiological appearance of roots, stems, leaves and flowers
Tray A normal watering				
Tray B overwatering				

1 Calculate the percentage germination from each treatment.
2 Describe any physiological differences between the two groups of seedlings.

Conclusions

1 Explain the differences in your observations.
2 Identify any diseases present on the waterlogged specimens.
3 Name another disease common in waterlogged conditions.

Exercise 12.3 *Capillary rise*

Background

Water moves through the spaces (pores) between soil particles. This is called capillary action. As a plant absorbs water through the roots it creates a space into which other water will flow rather like a siphon. Water may also evaporate from the surface of the soil drawing more water upwards from the subsoil also by capillary action. The size of the soil pores will influence the speed with which capillary action works and also the ability of the soil to drain effectively.

Aim

To demonstrate the process of capillary rise (water movement through soil).

Apparatus

    blended sand grades (3)
    capillary tubes (3)
    retort stand
    tray
    water

Method

You may find it helpful to draw a diagram of the apparatus.

1 Fill each tube with a different grade of sand.
2 Clamp each tube in a retort stand.
3 Place tubes in a tray of water.
4 Observe the reaction.

Results

1 Measure the water level reached in each tube and record your result in the table provided.
2 Explain the variations in the readings obtained.

Capillary tube	Height of wetting front (mm)	Explanation of variations
Tube A		
Tube B		
Tube C		

Conclusions

1 What process is responsible for the rise in the level of the water in the three tubes?
2 Define the term 'capillary rise'.
3 What significance are your findings for drainage of horticultural land?
4 Describe the significance of your finding for the irrigation of plants.
5 Describe the influence of capillary rise on the watering of container-grown plants.

Exercise 12.4 *Plants' physiological responses to water stress*

Background

Plants will vary in their response to different degrees of water stress. Waterlogged soils tend to produce shrivelled leaves and roots. The rotten egg smell associated with hydrogen sulphide gas produced in anaerobic conditions is also a common symptom. Drought conditions often result in a flush of root growth as plants actively seek out new sources of water, followed by leaf fall and plant death. Usually, plants that are semi-submerged in water respond by their roots growing upwards away from the waterlogged conditions. Container-grown house plants that have been overwatered often demonstrate these characteristics. This is because aerobic conditions are essential for healthy growth and development. This exercise will investigate some of these characteristics.

Aim

To observe the effects of: (a) waterlogging the root zone, (b) half waterlogging the root zone, and (c) normal watering.

Apparatus

> pelargoniums pot plants (3)
> reservoirs (e.g. bucket) (3)
> water

Method

1 Stand plant A in a reservoir with the water completely submersing the root zone (waterlogged).
2 Stand plant B in a reservoir with water filled to half the depth of the pot.
3 Stand plant C in a reservoir and water normally.
4 Observe each specimen over the next 7 days.
5 Examine the leaves, stems, flowers and root balls.
6 State how their outward appearance varies.

Results

Enter results in the table provided.

Treatment	Response to water stress			
	Roots	Stem	Leaves	Flowers
Plant A (waterlogged)				
Plant B (semi-waterlogged)				
Plant C (normal irrigation)				

Conclusions

1 Describe the physiological reaction of plants to waterlogging.
2 Explain the conditions that favour the production of hydrogen sulphide gas.
3 Give two management practices in horticulture that might lead to semi-waterlogged conditions.

13 Measuring soil pH

Background

The importance of pH

pH is used to measure the acidity or alkalinity of the growth medium. Plants grown in soils with inappropriate pH will fail. For example, the public often purchase heathers and other acid-loving species, plant them in alkaline soils and complain when they grow poorly or die. In addition, the availability of nutrients changes with pH, becoming either deficient or toxic in the extremes. The pH at which most nutrients are freely available for absorption by plant roots is pH 6.5 in soils and pH 5.8 in composts. Similarly, there are many plants grown in commercial crop production that will only tolerate a narrow band of acidity. Accurate pH management is therefore vital to ensure success.

As horticulturalists we need to know:

- the importance of pH
- methods of measuring soil pH
- specific plant pH preferences
- methods to correct soil acidity by liming (to raise pH)
- methods to increase soil acidity by sulphur (to lower pH).

The pH scale

pH is measured on a log scale from 0 to 14, as illustrated in Figure 13.1. pH 7.0 is neutral (neither acid or alkaline); below pH 7.0 is acidic and above is alkaline.

Plants vary greatly in their tolerance to acidity levels. Lime-loving (alkaline-loving) plants are called **calcicole** species.

The pH scale. pH is measured on a **log scale** from 0–14:

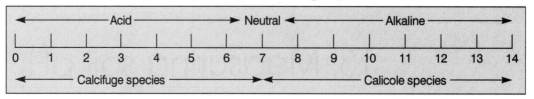

Figure 13.1
The pH scale

Lime-hating (acid-loving) plants, are called **calcifuge** species. A list of plant pH preferences is given in Table 13.2.

pH and hydrogen ion (H⁺) concentration

Technically, pH is defined as the negative logarithm of the hydrogen ion concentration. The significance of this log scale is that, for example, pH 5 is ten times more acidic than pH 6, but one hundred times more acidic than pH 7, and will therefore require substantially more lime to raise the pH. Table 13.1 may help to explain the relationship between pH and some common materials.

Table 13.1
pH and H⁺ concentration, with examples

H^+ concentration	Soil reaction	pH	Example substance
10^{-0}	pure acids	0	sulphuric acid
10^{-1}	extremely acid	1	battery acid, gastric juices
10^{-2}	extremely acid	2	worst acid rain record
10^{-3}	very strongly acid	3	coke, lemon, vinegar, sweat
10^{-4}	strongly acid	4	orange, lemonade
10^{-5}	moderately acid	5	urine
10^{-6}	slightly acid	6	clean rain, coffee
10^{-7}	**neutral**	**7**	**distilled water**
10^{-8}	slightly alkaline	8	sea water, baking soda
10^{-9}	moderately alkaline	9	soap solution
10^{-10}	strongly alkaline	10	milk of magnesia
10^{-11}	very strongly alkaline	11	strong bleach
10^{-12}	extremely alkaline	12	ammonia smelling salts
10^{-13}	extremely alkaline	13	caustic soda solution
10^{-14}	pure alkaline	14	concentrated sodium hydroxide

Table 13.2
Plant tolerance

Name	pH range	Name	pH range	Name	pH range

Loam and mineral soil mediums

Nursery stock (including ornamental trees, flowers and shrubs)

Name	pH range	Name	pH range	Name	pH range
Abelia	6.0–8.0	Colchicum	5.5–6.5	Iris	5.0–6.5
Acacia	6.0–8.0	Columbine	6.0–7.0	Ivy	6.0–8.0
Acanthus	6.0–7.0	Convolvulus	6.0–8.0		
Acer	5.5–6.5	Coreopsis	5.0–6.0	Juniper	5.0–6.5
Aconitum	5.0–6.0	Coronilla	6.5–7.5		
Adonis	6.0–8.0	Corydalis	6.0–8.0	Kalmia	4.5–5.0
Ageratum	6.0–7.5	Cosmos	5.0–8.0	Kerria	6.0–7.0
Ailanthus	6.0–7.5	Cotoneaster	6.0–8.0		
Ajuga	4.0–6.0	Crab apple	6.0–7.5	Laburnum	6.0–7.0
Althea	6.0–7.5	Crocus	6.0–8.0	Larch	4.5–7.5
Alyssum	6.0–7.5	Cynoglossum	6.0–7.5	Laurel	4.5–6.0
Amaranthus	6.0–7.5			Lavender	6.5–7.5
Anchusa	6.0–7.5	Daffodil	6.0–6.5	Liatris	5.5–7.5
Androsace	5.0–6.0	Dahlia	6.0–7.5	Ligustrum	5.0–7.5
Anemone	6.0–7.5	Day lily	6.0–8.0	ovalifolium	6.5–7.0
Anthyllis	5.0–6.0	Delphinium	6.0–7.5	Lilac	6.0–7.5
Arbutus	4.0–6.0	Deutzia	6.0–7.5	Lily-of-the-valley	4.5–6.0
Ardisia	6.0–8.0	Dianthus	6.0–7.5	Lithospermum	5.0–6.5
Arenaria	6.0–8.0	Dogwood	5.0–6.5	Lobelia	6.0–7.5
Aristea	6.0–7.5			Lupinus	5.5–7.0
Armeria	6.0–7.5	Elaeagnus	5.0–7.5		
Arnica	5.0–6.5	Enkianthus	5.0–6.0	Magnolia	5.0–6.0
Ashodoline	6.0–8.0	Erica		Mahonia	6.0–7.0
Asperula	6.0–8.0	carnea	4.5–5.5	Marguerite	6.0–7.5
Aster	5.5–7.5	cinerea	5.5–6.5	Marigold	5.5–7.0
Astilbe	6.0–8.0	Eucalyptus	6.0–8.0	Molinia	4.0–5.0
Aubrieta	6.0–7.5	Euphorbia	6.0–7.0	Moraea	5.5–6.5
Azalea	4.5–6.0	Everlasting flowers	5.0–6.0	Morning glory	6.0–7.5
		Fir	4.0–6.5	Moss	6.0–8.0
Beauty bush	6.0–7.5	Firethorn	6.0–8.0	sphagnum	3.5–5.0
Bergenia	6.0–7.5	Forget-me-not	6.0–8.0	Mulberry	6.0–7.5
Betula papyrifera	5.5–6.5	Forsythia	6.0–8.0		
Bleeding heart	6.0–7.5	Foxglove	6.0–7.5	Narcissus	6.0–7.5
Bluebell	6.0–7.5	Fritillaria	6.0–7.5	Nasturtium	5.5–7.5
Broom	5.0–6.0	Fuchsia	5.5–6.5	Nicotiana	5.5–6.5
Buddleia	6.0–7.0				
Buphthalum	6.0–8.0	Gaillardia	6.0–7.5	Pachysandra	5.0–8.0
		Gardenia	5.0–7.0	Paeonia	6.0–7.5
		Gazania	5.5–7.0	Pansy	5.5–7.0
Calendula	5.5–7.0	Gentiana	5.0–7.5	Pasque flower	5.0–6.0
Calluna vulgaris	4.5–5.5	Geum (Avens)	6.0–7.5	Passion flower	6.0–8.0
Camassia	6.0–8.0	Ginkgo	5.5–7.5	Paulownia	6.0–8.0
Camellia	4.5–6.0	Gladiolus	6.0–7.0	Picea pungens	4.5–5.5
Candytuft	6.0–7.5	Globularia	5.5–7.0	Pea, sweet	6.0–7.5
Canna	6.0–8.0	Godetia	6.0–7.5	Pelargonium	6.0–7.5
Canterbury bells	6.0–7.5	Goldenrod	5.0–7.0	Penstemon	5.5–7.0
Carnation	6.0–7.5	Gourd	6.0–7.0	Peony	6.0–7.5
Catalpa	6.0–8.0	Gypsophila	6.0–7.5	Periwinkle	6.0–7.5
Celosia	6.0–7.0			Pittosporum	5.5–6.5
Centaurea	5.0–6.5	Hawthorn	6.0–7.0	Plantain	6.0–7.5
Cerastium	6.0–7.0	Helleborus	6.0–7.5	Polyanthus	6.0–7.5
Chamaecyparis		Hibiscus	6.0–8.0	Poplar	5.5–7.5
lawsoniana		Holly	5.0–6.5	Poppy	6.0–7.5
'Columnaris'	4.5–5.5	Hollyhock	6.0–8.0	Portulaca	5.5–7.5
Chrysanthemum	6.0–7.0	Horse-chestnut	5.5–7.0	Primrose	5.5–6.5
Cissus	6.0–7.5	Hydrangea		Privet	6.0–7.0
Cistus	6.0–7.5	blue	4.0–5.0	Prunella	6.0–7.5
Clarkia	6.0–6.5	pink	6.0–7.0	Prunus	6.5–7.5
Clematis	5.5–7.0	white	6.5–8.0	Pyracantha	5.0–6.0
Clianthus	6.0–7.5	Hypericum	5.5–7.0		

Table 13.2
Plant tolerance (continued)

Name	pH range	Name	pH range	Name	pH range
Nursery stock (including ornamental trees, flowers and shrubs) – continued					
Quercus (oak)	5.5–6.5	Sea grape	5.0–6.5	Tamarix	6.5–8.0
		Sedum	6.0–8.0	Tobacco	5.5–7.5
Red hot poker	6.0–7.5	Snapdragon	5.5–7.0	Trillium	5.0–6.5
Rhododendron	4.5–6.0	Snowdrop	6.0–8.0	Tulip	6.0–7.0
Rosa laxa	6.5–7.0	Soapwort	6.0–7.5		
Rosa multiflora	5.5–6.5	Speedwell	5.5–6.5	Viburnum	5.0–7.5
Rose		Spiraea	6.0–7.5	Violet	5.0–7.5
climbing	6.0–7.0	Spruce	4.0–5.9	Virginia creeper	5.0–7.5
hybrid tea	5.5–7.0	Stock	6.0–7.5		
rambling	5.5–7.0	Stonecrop	6.5–7.5	Wallflower	5.5–7.5
Rowan	4.5–6.5	Sumach	5.0–6.5	Water lily	5.5–6.5
		Sunflower	6.0–7.5	Weigela	6.0–7.0
Saintpaulia	6.0–7.0	Sweet William	6.0–7.5	Wisteria	6.0–8.0
Salix	6.5–7.0	Sycamore	5.5–7.5		
Salvia	6.0–7.5	Syringa	6.0–8.0	Yew	5.0–7.5
Scabiosa	5.0–7.5	vulgaris	6.5–7.0	Zinnia	5.5–7.5
Grasses					
Annual meadow grass	5.5–7.5	Clover	6.0–7.0	Pampas	6.0–8.0
Bents	5.5–6.5	Cocksfoot	5.3–7.5	Rye	5.8–7.4
Browntop	4.5–6.5	Crested dogs tail	5.0–6.5		
Colonial	5.6–7.0			Smooth-stalked	
Creeping	5.5–7.5			meadow grass	6.0–8.0
Velvet	5.2–6.5	Fescues		Timothy	5.3–7.0
Bermuda grass	6.0–7.0	Chewing	4.5–8.0	Trefoil	6.1–7.5
Bluegrass		Hard	4.0–5.5		
Annual	5.5–7.0	Meadow	6.0–7.5	Vetches	5.9–7.0
Canada	5.7–7.2	Red	4.5–8.0	Wheat grass	6.1–8.6
Kentucky	5.8–7.5	Sheep's	4.0–5.5	Yorkshire fog	4.6–7.0
Rough	5.8–7.2	Tall	5.5–7.0		
Common weeds (wild plants)					
Annual meadow grass	5.5–7.5	Creeping soft grass	4.6*	Oxeye daisy	6.1*
		Cowslip	5.1*	Rest harrow	6.1*
Bird's foot trefoil	6.1*	Dog violet	5.1*		
Burnet	6.1*	Gentian	6.1*	Wild carrot	5.6*
				Wild clematis	6.1*
Cleavers	6.1*	Kidney vetch	6.1*	Woodrush	4.6*
Colt's foot	5.6*				
Common nettle	5.1*	Milkwort	6.1*	Yorkshire fog	4.6*
Fruit crops					
Apple	5.0–6.5	Damson	6.0–7.5	Nectarine	6.0–7.5
Apricot	6.0–7.0				
Avocado	6.0–7.5	Gooseberry	5.0–6.5	Pawpaw	6.0–7.5
		Grapevine	6.0–7.5	Peach	6.0–7.5
Banana	5.0–7.0			Pear	5.3–7.5
Blackberry	4.9–6.0	Hazelnut	6.0–7.0	Pineapple	5.0–6.0
Blueberry	4.5–6.0	Hop	6.0–7.5	Plum	5.6–7.5
				Pomegranate	5.5–6.5
Cherry	6.0–7.5	Lemon	6.0–7.0		
Cranberry	4.0–5.5	Lychee	6.0–7.0	Quince	6.0–7.5
Currants					
black	6.0–8.0	Mango	5.0–6.0	Raspberry	5.5–6.5
red	5.5–7.0	Melon	5.5–6.5	Rhubarb	5.5–7.0
white	6.0–8.0	Mulberry	6.0–7.5	Strawberry	5.1–7.5

* Threshold below which plant growth is known to suffer.

Table 13.2
Plant tolerance (continued)

Vegetables, herbs and some arable crops

Name	pH range	Name	pH range	Name	pH range
Artichoke	6.5–7.5	Fennel	5.0–6.0	Peanut	5.0–6.5
Asparagus	5.9–8.0			Pepper	5.5–7.0
		Garlic	5.5–7.5	Peppermint	6.0–7.5
Barley	5.9–7.5	Ginger	6.0–8.0	Pistachio	5.0–6.0
Basil	5.5–6.5	Ground nut	5.5–6.5	Potato	4.9–6.0
Beans				Potato, sweet	5.5–6.0
broad	6.0–7.5	Horseradish	6.0–7.0	Pumpkin	5.5–7.5
french	6.0–7.5	Kale	5.4–7.5		
runner	6.0–7.5	Kohlrabi	6.0–7.5	Radish	6.0–7.0
Beet				Rape	5.6–7.5
sugar	5.9–7.5	Leek	5.8–8.0	Rice	5.0–6.5
table	5.9–7.5	Lentil	5.5–7.0	Rosemary	5.0–6.0
Beetroot	6.0–7.5	Lettuce	6.1–7.0	Rhubarb	5.4–7.0
Broccoli	6.0–7.0	Linseed	5.4–7.5		
Brussels sprouts	5.7–7.5			Sage	5.5–6.5
		Marjoram	6.0–8.0	Shallot	5.5–7.0
Cabbage	5.4–7.5	Marrow	6.0–7.5	Sorghum	5.5–7.5
Calabrese	6.5–7.5	Millet	6.0–6.5	Soyabean	5.5–6.5
Cantaloupe	6.0–8.0	Mint	6.6–8.0	Spearmint	5.8–7.5
Carrots	5.7–7.0	Mushroom	6.5–7.5	Spinach	6.0–7.5
Cauliflower	5.5–7.5	Mustard	5.4–7.5	Sugar-cane	6.0–8.0
Celery	6.3–7.0			Swede	5.4–7.5
Chicory	5.1–6.5	Oats	5.4–7.5		
Chinese leaf	6.0–7.5	Olive	5.5–6.5	Thyme	5.5–7.0
Corn, sweet	5.5–7.5	Onion	5.7–7.0	Tomato	5.1–7.5
Cotton	5.0–6.0			Turnip	5.4–7.0
Courgettes	5.5–7.0	Paprika	7.0–8.5		
Cowpea	5.0–6.5	Parsley	5.1–7.0	Watercress	6.0–8.0
Cress	6.0–7.0	Parsnip	5.4–7.5	Watermelon	5.0–6.5
Cucumber	5.5–7.0	Pea	5.9–7.5	Wheat	5.5–7.5

Loamless compost

Container-grown nursery stock and protected pot plants (house plants)

Name	pH range	Name	pH range	Name	pH range
Bedding plants	5.5–6.5	Blood leaf	5.5–6.5	Coffee plant	5.0–6.0
Ericaeous	5.0–5.5	Bottle brush	6.0–7.5	Coleus	6.0–7.0
General plants	5.5–6.5	Bougainvillea	5.5–7.5	Columnea	4.5–5.5
Seed compost	5.5–6.5	Bromeliads	5.0–6.0	Coral berry	5.5–7.5
		Butterfly flower	6.0–7.5	Crassula	5.0–6.0
Abutilon	5.5–6.5			Creeping fig	5.0–6.0
Acorus	5.0–6.5	Cactus	4.5–6.0	Croton	5.0–6.0
Aechmea	5.0–5.5	Caladium	6.0–7.5	Crown of thorns	6.0–7.5
African violet	6.0–7.0	Calceolaria	6.0–7.0	Cuphea	6.0–7.5
Aglaonema	5.0–6.0	Calla lily	6.0–7.0	Cyclamen	6.0–7.0
Amaryllis	5.5–6.5	Camellia	4.5–5.5	Cyperus	5.0–7.5
Anthurium	5.0–6.0	Campanula	5.5–6.5		
Aphelandra	5.0–6.0	Capsicum pepper	5.0–6.5	Daphne	6.5–7.5
Aralia	6.0–7.5	Cardinal flower	5.0–6.0	Dieffenbachia	5.0–6.0
Araucaria	5.0–6.0	Carnations (pinks)	5.5–6.5	Dipladenia	6.0–7.5
Asparagus fern	6.0–8.0	Castor oil plant	5.5–6.5	Dizygotheca	6.0–7.5
Aspen	4.0–5.5	Century plant	5.0–6.5	Dracaena	5.0–6.0
Aspidistra	4.0–5.5	Chinese primrose	6.0–7.5	Dragon tree	5.0–7.5
Azalea	4.5–6.0	Christmas cactus	5.0–6.5	Dutchman's pipe	6.0–8.0
		Chrysanthemum	5.5–6.5		
Begonia	5.5–7.0	Cineraria	5.5–7.0	Easter lily	6.0–7.0
Bird of paradise	6.0–6.5	Clerodendrum	5.0–6.0	Elephant's ear	6.0–7.5
Bishop's cap	5.0–6.0	Clivia	5.5–6.5	Episcia	6.0–7.0
Black-eyed-Susan	5.5–7.5	Cockscomb	6.0–7.5	Euonymus	5.5–7.0

Table 13.2
Plant tolerance (continued)

Name	pH range	Name	pH range	Name	pH range
Container-grown nursery stock and protected pot plants (house plants) – continued					
Feijoa	5.0–7.5	Jacaranda	6.0–7.5	Pitcher plant	4.0–5.5
Fern		Japanese sedge	6.0–8.0	Plumbago	5.5–6.5
bird's nest	5.0–5.5	Jasminem	5.5–7.0	Podocarpus	5.0–6.5
Boston	5.5–6.5	Jerusalem cherry	5.5–6.5	Poinsettia	6.0–7.5
Christmas	6.0–7.5			Polyscias	6.0–7.5
cloak	6.0–7.5	Kaffir fig	6.0–7.5	Pothos	5.0–6.0
feather	5.5–7.5	Kalanchoe	6.0–7.5	Prayer plant	5.0–6.0
hart's tongue	7.0–8.0	Kangaroo thorn	6.0–8.0	Punica	5.5–6.5
holly	4.5–6.0	Kangaroo vine	5.0–6.5		
maidenhair	6.0–8.0			Rubber plant	5.0–6.0
rabbit's foot	6.0–7.5	Lace flower	6.0–7.5		
spleenwort	6.0–7.5	Lantana	5.5–7.0	Sansevieria	4.5–7.0
Fig	5.0–6.0	Laurus (bay tree)	5.0–6.0	Saxifraga	6.0–8.0
Fittonia	5.5–6.5	Lemon plant	6.0–7.5	Schizanthus	6.0–7.0
Freesia	5.5–6.5			Scilla	6.0–8.0
French marigold	5.0–7.5	Mimosa	5.0–7.0	Scindapsus	5.0–6.0
		Mind-your-own-business	5.0–5.5	Selaginella	6.0–7.0
Gardenia	5.0–6.0	Monstera	5.0–6.0	Senecio	6.0–7.0
Genista	6.5–7.5	Myrtle	6.0–8.0	Shrimp plant	5.5–6.5
Geranium	6.0–8.0			Spanish bayonet	5.5–7.0
Gloxinia	5.5–6.5	Nephthytis (syngonium)	4.5–5.5	Spider plant	6.0–7.5
Grape hyacinth	6.0–7.5	Never never plant	5.0–6.0	Succulents	5.0–6.5
Grape ivy	5.0–6.5			Syngonium	5.0–6.0
Grevillea	5.5–6.5	Oleander	6.0–7.5		
Gynura	5.5–6.5	Oplismenus	5.0–6.0	Thunbergia	5.5–7.5
		Orange plant	6.0–7.5	Tolmiea	5.0–6.0
Hedera (ivy)	6.0–7.5	Orchid	4.5–5.5	Tomato	5.5–6.0
Heliotropium	6.0–8.0	Oxalis	6.0–8.0	Tradescantia	5.0–6.0
Helxine	5.0–6.0				
Herringbone plant		Painted lady	6.0–7.5	Umbrella tree	5.0–7.5
(prayer plant)	5.0–6.0	Palms	6.0–7.5		
Hibiscus	6.0–8.0	Pandanus	5.0–6.0	Venus flytrap	4.0–5.0
Hoya	5.0–6.5	Patient Lucy	5.5–6.5	Verbena	6.0–8.0
Hyacinth	6.5–7.5	Peacock plant	5.0–6.0	Vinca	6.0–7.5
		Pelargoniums	6.0–7.5		
Impatiens	5.5–6.5	Pellionia	5.0–6.0	Weeping fig	5.0–6.0
Indigofera	6.0–7.5	Philodendron	5.0–6.0		
Iresine	5.0–6.5	Phlox	5.0–6.5	Yucca	6.0–8.0
Ivy tree	6.0–7.0	Pilea	6.0–8.0	Zebrina	5.0–6.0

pH range and plant tolerance

The list in Table 13.2 gives the pH tolerance range for the most common amenity and commercial crop production species. The optimum pH would normally be in the middle of the range. Where only a lower figure is given, this is the threshold below which plant growth is known to suffer (shown by an asterisk).

Exercise 13.1 *Soil and compost pH testing*

Background

Before we can modify growth medium pH it is necessary to determine the existing level of soil acidity. Comparison can then be made between this and the plant pH tolerance range for the species growing. If the acidity level is too low, it will need to be corrected by liming. If it is too high, it will need to be lowered, normally by applying sulphur.

There are several methods available to determine pH levels in the growth medium. These range from litmus indicator paper, garden centre-type needle probe (e.g. Rapitest), test tube measurements (e.g. BDH method), portable electrode meters (e.g. Hanna Instruments pHep range) and laboratory-based analysis (e.g. ADAS or local college services).

Of these, the test tube method is most suitable for determining quickly, on site, the pH of soils and composts. The BDH method will be considered in this exercise. Several products are available that use similar principles including Rapitest and Sudbury products; ADAS field kits and many amateur kits are available from garden centres.

The BDH method for pH testing

British Drug House (BDH, now owned by Merkoquant) developed the apparatus for the following technique to determine soil pH. It involves mixing a small amount of soil with a white powder called barium sulphate, some distilled water and universal indicator solution. The cocktail is then shaken up and left to stand. The barium sulphate causes the soil to fall to the bottom of the test tube (flocculate). It is a neutral material, and like distilled water does not influence the pH. After a few minutes the soil particles will settle out, leaving a coloured solution above. This coloured solution is then compared with a colour chart and the corresponding pH recorded.

Figure 13.2
Soil and compost pH tests

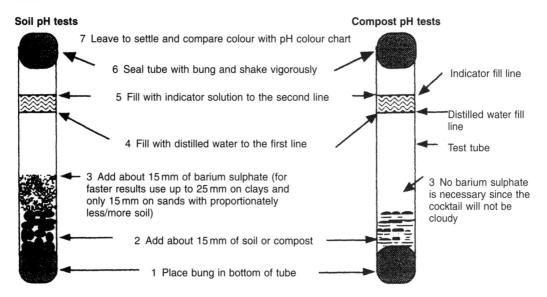

Soil pH tests

Compost pH tests

7 Leave to settle and compare colour with pH colour chart

6 Seal tube with bung and shake vigorously

Indicator fill line

5 Fill with indicator solution to the second line

Distilled water fill line

4 Fill with distilled water to the first line

Test tube

3 Add about 15 mm of barium sulphate (for faster results use up to 25 mm on clays and only 15 mm on sands with proportionately less/more soil)

3 No barium sulphate is necessary since the cocktail will not be cloudy

2 Add about 15 mm of soil or compost

1 Place bung in bottom of tube

Aim

To determine the pH of soil and compost growth mediums using the BDH method.

Apparatus

BDH test tubes	soil A
barium sulphate	soil B
universal indicator solution	compost A
distilled water	compost B
pH colour chart	

Method

Follow the method in Figure 13.2 to find the pH of the different materials.

Results

Record your results in the table provided.

Growth medium	pH	Interpretation (e.g. slightly acid)
Soil A		
Soil B		
Compost A		
Compost B		

Conclusions

1 Explain the purpose of using barium sulphate.
2 State the ideal pH for most plants grown in a soil.
3 Comment on the suitability of soil A for growing white-flowered hydrangeas.
4 How does the method vary, if testing a compost?
5 What is the ideal pH for most plants grown in a compost?
6 Comment on the suitability of soil B for growing strawberries.
7 Comment on the suitability of compost A for growing azaleas.
8 Comment on the suitability of compost B for growing busy lizzies (Impatiens).
9 Explain what is meant by a 'calcicole plant' and give one example.
10 Explain what is meant by a 'calcifuge plant' and give one example.

Exercise 13.2 *Soil texture assessment*

Background

In addition to determining the soil pH information about the soil texture is also needed. Soil texture refers to the relative proportions of sand, silt and clay particles in the soil. This is important because clay soils require more lime to correct acidity than sandy soils. This property is called buffering capacity.

Please refer to Chapter 10 for a detailed procedure for assessing soil texture.

Aim

To correctly conduct a hand texture assessment of different soil types.

Apparatus

Figure 10.3
soil A
soil B

Method

Assess the texture of soil A and soil B, giving the reasoning behind your answer.

Results

Record your results in the table provided.

Sample	Texture	Reasoning
Soil A		
Soil B		

Conclusions

1 Explain which soil has a greater buffering capacity and also therefore ability to hold nutrients.
2 Which soil will require more lime to correct any soil acidity problem, and why?
3 How would the procedure to measure pH, using the BDH method, vary between these two soils?

14 Raising soil pH

Background

Over a period of years the pH of the growth medium will naturally tend to become more acidic. This is due to a variety of processes. For example, rainfall washes out (leaches) free lime (calcium carbonate $CaCO_3$) from the soil and in addition acid rainfall builds up the hydrogen ion concentration. Some plants, such as ericaceous and conifer species, acidify the soil due to organic acids being leached from their leaves. Micro-organism activity may vary pH by up to one pH unit over the course of a year, due to the gases given off during respiration and other chemical reactions.

Once a soil acidity problem has been identified action will need to be taken to correct it. Any material used to correct soil acidity is known as **lime**. The amount of lime needed will be determined by reference to:

- present pH
- soil texture and buffering capacity
- effectiveness of the liming material (neutralizing value).

Exercise 14.1 *Soil texture and lime requirements*

Background

Soil texture greatly influences the amount of lime that is required to raise pH. It is more difficult to alter the pH of a clay soil than of a sandy soil due to different buffering capacities.

The amount of lime required to change pH is given in Table 14.1. By convention this is always stated using ground limestone, in tonnes per hectare. This can be converted to grammes per metre square, by multiplying by 100. The figures can easily be plotted into a graph and may make it helpful for future reference.

Present pH	Loamy sands	Sandy loams	Silt loams	Loams	Clay loams
3.5	15.0	18.1	21.0	24.0	31.0
4.0	12.5	15.0	17.5	20.0	25.0
4.5	10.0	11.9	14.0	16.2	20.0
5.0	7.5	9.4	10.0	11.9	15.0
5.5	5.0	6.2	7.0	8.1	10.0
6.0	2.5	3.1	3.5	3.7	5.0

Table 14.1
Lime required to raise pH to 6.5

Tonnes/hectare of ground limestone (multiply by 100 for g/m^2).

Aim

To gain competence at gathering and interpreting information on soil texture and liming requirements to correct soil acidity.

Apparatus

Table 14.1
Example situations

Method

Either using Table 14.1, or a graph plotted from the data, state the liming requirement to raise pH to pH 6.5 under the conditions given below.

Results

Enter your results in the table provided.

State the amount of lime required to raise the pH of	Lime requirement (ground limestone)	
	t/ha	g/m^2
1 A clay loam cricket square from pH 4.5 to pH 6.5		
2 A clay bedding plant area from pH 5.0 to pH 6.5		
3 A loamy sand old lawn from pH 5.5 to pH 6.5		
4 A rugby pitch (silt loam) from pH 4.5 to pH 6.5		
5 Allotments (loams) from pH 5.0 to pH 6.5		
6 An unused acid bed (sandy loam) from pH 4.0 to pH 6.5		

Conclusions

1 Explain how the pH of soil would naturally become more acidic.
2 Which sites have the greatest buffering capacity?
3 What is the effect of an inappropriate level of soil acidity on plant growth?
4 Explain what is meant by the term 'lime'.

Exercise 14.2 *Liming materials and neutralizing values*

Background

Any material used to correct soil acidity is called lime. A lime requirement is the amount of lime required to raise the top 150 mm of the soil to pH 6.5. For the best results finely ground lime should be applied several months before planting.

The application of liming materials generally corrects soil acidity by neutralizing and replacing hydrogen ions with calcium ions. There are many different forms of lime, all of which have different calcium ion contents. Because of this, they will vary in their effectiveness to correct soil acidity. Limes with a high calcium ion content will be more effective than limes with a low calcium ion content and will therefore require less material to neutralize acidity.

We can compare the effectiveness of different liming materials by referring to their neutralizing values (NV). All liming materials are compared with pure calcium oxide (CaO). The neutralizing value compares how

Material	Neutralizing value
burnt lime (quicklime, calcium oxide)	85– 90
burnt magnesian limestone	95–110
calcareous shell sand	24– 45
dolomitic limestone	48
ground chalk	48– 50
ground limestone	48– 50
ground magnesian limestone	50– 55
hydrated lime (slaked lime)	70
marl	variable
pure calcium carbonate	56
pure calcium hydroxide	74
pure calcium oxide	**100**
screened chalk	45
waste limes (basic slag)	variable

Table 14.2
Common liming materials

effective the liming material is compared with using pure calcium oxide. For example, if using ground limestone, twice as much material would be needed than if using pure calcium oxide. Ground limestone therefore has a neutralizing value of 50, or is only 50 per cent as effective in correcting soil acidity as pure calcium oxide. Pure calcium oxide has a neutralizing value of 100. The neutralizing values may vary slightly between batches, but will always be stated on the bag in which the lime is sold. A list of some common liming materials and their neutralizing values is given in Table 14.2.

A lime requirement is always stated in terms of ground limestone (in t/ha). However, it may be more appropriate to use a cheaper or more convenient alternative liming material and by comparing their neutralizing values an equivalent application rate can be found.

If we know the quantity of ground limestone required to raise the pH to 6.5, and we know the neutralizing value of the alternative material, we can calculate how much of the alternative liming material is needed, using the following formula:

$$\text{Lime requirement of alternative material (t/ha)} = \text{Quantity of ground limestone recommended} \times \frac{\text{NV ground limestone}}{\text{NV alternative material}}$$

For example, an application rate of 2.5 t/ha ground limestone (NV 50), is recommended to correct soil acidity on a loamy sand land reclamation site. It is more convenient to use a local supplier of screened chalk (NV 45). Following the above equation the conversion is as follows:

$$\text{Amount of screened chalk (NV 45) required} = \frac{2.5 \text{ t/ha ground limestone}}{} \times \frac{50 \text{ (NV ground limestone)}}{45 \text{ (NV screened chalk)}}$$

$$= 2.5 \times \frac{50}{45}$$

$$= 2.78 \text{ t/ha screened chalk}$$

Aim

To develop skills at calculating the quantity of alternative liming materials required to correct soil acidity and raise pH to 6.5.

Apparatus

Table 14.1
Table 14.2
sample situations given below

Method

Using the table of liming materials and the formula provided, calculate an alternative liming requirement for the situations given below.

Results

1 An application rate of 10 t/ha ground limestone (NV 50) is recommended to raise the pH from 5.5 to 6.5 on the amenity beds (clay loam) behind a housing development. What is the alternative requirement for:

(a) calcareous shell sand (NV 25)?
(b) hydrated lime (NV 70)?

2 An application rate of 14 t/ha ground limestone (NV 50) is recommended to raise the pH from 4.5 to 6.5 on an unused acid bed (silt loam texture). What is the alternative requirement for:

(a) calcareous shell sand (NV 25)?
(b) screened chalk (NV 45)?

3 An application rate of 2.5 t/ha ground limestone (NV 50) is recommended to raise the pH from 6.0 to 6.5 on a utility lawn (loamy sand texture). What is the alternative requirement for:

(a) calcareous shell sand (NV 25)?
(b) pure calcium oxide (NV 100)?

Conclusions

1 Which liming material is most commonly available in garden centres?
2 What information about the liming material needs to be recorded from every packet of lime?
3 How is t/ha converted to g/m^2?
4 When should lime 'ideally' be applied?

Exercise 14.3 *Integration of skills*

Background

Sample soil A and sample soil B are both taken from recreational areas in different parts of the county. The growth of plants has been very poor and this is thought to be due to an acid soil pH. Determine whether or not this is true, and if so, produce a liming recommendation to correct the soil acidity to pH 6.5. However, as an additional consideration, the County Council Parks Department has stated that they only stock calcareous shell sand (NV 25), screened chalk (NV 45), and hydrated lime (NV 70). Your recommendations must therefore be for these liming materials.

Aim

To use all the skills developed in these exercises to calculate a lime requirement for soil A and soil B.

Method

Investigate the soil samples and complete the table provided.

Results

Enter your results in the table provided.

Soil sample	Example soil	Soil A	Soil B
Present pH	4.5		
Texture	loamy sand		
Lime requirement (ground limestone, NV 50) (t/ha)	10.0		
Conversion to use calcareous shell sand (NV 25) (t/ha)	$10 \times \dfrac{50}{25} = 20$		
Conversion to use screened chalk (NV 45) (t/ha)	$10 \times \dfrac{50}{45} = 11$		
Conversion to use hydrated lime (NV 70) (t/ha)	$10 \times \dfrac{50}{70} = 7.14$		

Conclusions

1 Explain what is meant by the term 'lime'.
2 List six different liming materials.
3 How would you convert t/ha to g/m^2?

15 Lowering soil pH (increasing soil acidity)

Background

Soils that are too alkaline will require special corrective treatments and good management to achieve healthy plant growth. Problems of soil alkalinity normally occur due to an increasing proportion of sodium ions present in the soil solution. This may occur for a variety of reasons, such as:

- rising water tables redepositing previously leached ions (e.g. near flooding rivers)
- high dissolved nutrient content in irrigation water
- poor drainage, preventing leaching of dissolved ions
- poor watering practice, such as excessive drying out between irrigations.

Alkaline conditions cause the separation of soil clay particles and the breakdown of organic matter. Consequently, these soils may also have structural problems such as cracking in hot weather and becoming puddled when wet. Nitrogen and calcium deficiency are also common symptoms.

In other situations it may simply be that the intended plant's pH tolerance range is lower than the present soil pH and measures to lower pH (increase acidity) will be required (e.g. lowering pH of turfed areas to pH 6.0 to discourage weed growth).

Lowering the pH of the growth medium is a difficult and expensive process. It is normally achieved by the application of acid-forming sulphur substances, such as ground sulphur (rolls of sulphur), ferrous sulphate, aluminium sulphate, low grade pyrites and calcium sulphate (gypsum). The sulphur is changed to sulphuric acid by the action of soil micro-organisms and weathering processes. It can therefore cause plant death if applied to sites containing growing plants. Ideally it should be applied several months before planting. There are also several commercial products available that claim to lower the soil's pH over a period of months, for example. Phostrogen and Liquid Sod (flowable sulphur) products.

As a cheaper alternative, acid-based fertilizers can also help where only a slight adjustment is desired, possibly over a period of years. Likewise, sphagnum moss peat is naturally acidic (about pH 5.0) and this can be incorporated around the rootball for the rapid establishment of the plant (e.g. when growing ericaceous heathers in chalky soils).

The amount of sulphur needed to increase acidity is based on a number of existing soil properties. These are:

- existing soil pH
- soil texture
- free lime (calcium carbonate content).

The practice of measuring pH and assessing texture have been covered in Chapters 13 and 14, and the following exercises on these characteristics are offered for wholeness and practice.

Exercise 15.1 *Soil and compost pH testing*

Aim

To assess soil and compost pH using the BDH method.

Apparatus

2 alkaline soils labelled sample soil A and B
compost A (sphagnum moss peat)
compost B (ericaceous compost)
compost C (coir compost)
BDH pH testing kit

Method

Assess the pH of sample soils A and B, and sample composts A, B and C.

Results

Record your results in the table provided.

145

Growth medium	pH	Interpretation (e.g. slightly alkaline)
Sample soil A		
Sample soil B		
Compost A		
Compost B		
Compost C		

Conclusions

1 State which soil has the greatest alkalinity problem.
2 Explain how the alkalinity problem may have developed.
3 Explain how compost may be used to help rhododendrons establish on a slightly acid soil.

Exercise 15.2 *Soil texture*

Aim

To correctly conduct a hand texture assessment of sample soils A and B for the purpose of providing a subsequent recommendation to lower pH.

Apparatus

alkaline sample soil A and sample soil B
Figure 10.3
Table 13.2

Method

Assess the texture of sample soil A and sample soil B using the procedure for hand texture testing introduced in Chapter 10 (see Figure 10.3).

Results

Record your results in the table provided.

Sample	Texture	Reasoning
Sample soil A		
Sample soil B		

Conclusions

1 Which soil has a greater buffering capacity?
2 Which soil will require more sulphur to lower the pH, and why?
3 List three plant species that are suitable to be grown in these soils without altering the existing pH.

Exercise 15.3 *Free lime (calcium carbonate) content*

Background

Before we can effectively lower pH we must assess the free lime or calcium carbonate ($CaCO_3$) content of the soil and neutralize it through the addition of ground sulphur. This may not lower soil pH, merely neutralize the free lime. Once this has been done we can then apply additional sulphur to lower pH units further.

The free lime content is assessed by dropping a weak acid solution (10 per cent hydrochloric) onto the soil. The acid will neutralize some of the free lime and the reaction is observed by watching the size of the bubbles and listening to the sound of the fizz (hold close to ear). Although the industry standard is to use 10 per cent hydrochloric acid, alternative materials such as vinegar may be used. For each 0.1 per cent of free lime in the soil, 1 t/ha ground sulphur will be required to neutralize it. Table 15.1 states the quantities required and lists the reactions:

Table 15.1
Free lime content ($CaCO_3$) and neutralizing sulphur requirement

Description of free lime status	$CaCO_3$ content %	Reaction on contact with 10% hydrochloric acid		Ground sulphur required to neutralize free lime content (t/ha)
		Visible effects	Audible effects	
non-calcareous	0.1	none	none	1
very slightly calcareous	0.5	none	slightly audible	5
slightly calcareous	1.0	slight fizz on individual grains just visible	faintly audible	10
slightly calcareous	2.0	slightly more general fizz on close inspection	moderate–distinctly audible	20
calcareous	5.0	moderate fizz, bubbles up to 3 mm diameter	easily audible	50
very calcareous	10.0	general strong fizz; large bubbles up to 7 mm diameter easily seen	easily audible	100

Aim

To assess the free lime ($CaCO_3$) content and sulphur-neutralizing requirement of a range of horticultural materials.

Apparatus

sample soil A and sample soil B
5 types of sand and grit
petri dishes
vinegar
10 per cent hydrochloric acid solution (Warning – irritant and corrosive)

Method

1 Place a small amount of each sample material onto a petri dish.
2 Slowly add a few drops of 10 per cent hydrochloric acid (Warning – corrosive and irritant) onto the material and observe the reaction in accordance with Table 15.1 .
3 Replicate the experiment, this time using vinegar.
4 State the free lime content of the material measured by each acid.
5 Record the amount of sulphur required to neutralize the free lime content of the materials.

Results

Record your results in the table provided.

Horticultural material	Free Lime ($CaCO_3$) content		Ground sulphur required to neutralize free lime (t/ha)
	10% hydrochloric acid	vinegar	
sea shells			
beach sand			
grit			
silver sand			
horticultural sand			
sample soil A			
sample soil B			

Conclusions

1 What can you conclude about the use of vinegar as a substitute for 10 per cent hydrochloric acid?
2 Explain why sea shells are unsuitable for lowering soil pH.
3 Which sample soil has the greatest free lime content?
4 Explain why horticultural sand, used for container-grown plants, needs to be 'acid washed'.

Exercise 15.4 *Sulphur requirements to lower pH*

Background

Once the free lime has been neutralized, and for non-calcareous soils which have less than 0.1 per cent free lime, the quantities of sulphur required to lower the pH are dependent upon the soil texture and starting pH. Table 15.2 illustrates this relationship.

Refer to the soil texture exercises for the texture types that the labels represent in the table.

Table 15.2
Sulphur requirement to lower pH by one pH unit (t/ha)

Soil texture	S	LS	SL	SZL	ZL	SCL	CL	ZCL	SC	ZC	C
Sulphur requirement	0.75	1.0	1.0	1.5	1.8	2.2	2.2	2.2	2.7	2.7	2.7

Aim

To assess the quantity of sulphur required to lower soils of varying texture and pH.

Method

Using Table 15.2 answer the following questions.

Results

Enter your results in the table provided.

State the amount of ground sulphur required to lower the pH of:	Sulphur requirement (ground sulphur)	
	t/ha	g/m²
1 A seaside, sandy silt loam, bowling green from pH 7.0 to pH 6.0		
2 A waterlogged clay bedding plant area from pH 7.5 to pH 6.5		
3 A loamy sand planned rhododendron unit from pH 7.0 to pH 5.0		
4 An ericaceous unit (silt loam) from pH 6.5 to pH 5.0		
5 Allotments (silty clay loam) from pH 7.0 to pH 6.5		
6 An unused calicole species demonstration site (sandy loam) from pH 8.5 to pH 6.5		

Conclusion

State how much sulphur is required to lower the pH of sample soils A and B of Exercise 15.3 to pH 5.5.

Exercise 15.5 *Lowering pH – integration of skills*

Background

A poorly drained local common is being converted into a rhododendron and ericaceous heather park. Sample soils A and B, have been taken from this common for analysis. Calculate, using the skills developed during these exercises a sulphur requirement to lower the pH to pH 5.5.

Aim

To integrate and strengthen the skills developed in these exercises to calculate a sulphur requirement for a range of situations.

Method

Investigate soils A and B to complete the table provided.

Results

Enter your results in the table provided.

Growth medium	Soil texture	Free lime content CaCO₃ (%)	Sulphur required to neutralize free lime CaCO₃ (t/ha)	Present pH	Desired pH	Additional ground sulphur to lower pH (t/ha)	Total ground sulphur required (t/ha)
Example soil	clay loam	2.0	20	7.5	5.5	2.2 × 2 = 4.4	20 + 4.4 = 24.4
Sample soil A							
Sample soil B							

Conclusions

1 Why is it unwise to apply sulphur to areas where there are established plants growing?

2 How should sulphur be applied?

3 What other materials can be used to increase soil acidity in place of sulphur?

151

16 Standing beds in container production systems

Background

Container-grown plants have only a small water reserve available in the compost to meet their requirements. This soon becomes exhausted and needs to be regularly replenished through an irrigation system. Standing beds are an example of such a management tool in plant watering. The principles of operation are to manipulate the particle sizes to create a material that will allow even watering of plants in containers and yet have sufficient large pores to enable rapid drainage of excess rain water. Two categories can be distinguished:

- gravel beds
- sand beds

Gravel beds

Air enters the pores as water rapidly drains away. The capillary pathway between the container and the bed is broken and water in the compost cannot easily drain away. (See Figure 16.1.)

Sand beds

The sand bed becomes a part of the container system and a capillary pathway is maintained. In summer, water is drawn up

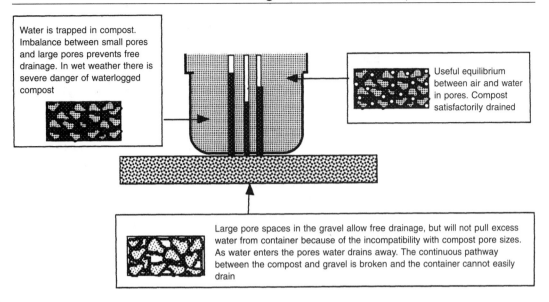

Water is trapped in compost. Imbalance between small pores and large pores prevents free drainage. In wet weather there is severe danger of waterlogged compost

Useful equilibrium between air and water in pores. Compost satisfactorily drained

Large pore spaces in the gravel allow free drainage, but will not pull excess water from container because of the incompatibility with compost pore sizes. As water enters the pores water drains away. The continuous pathway between the compost and gravel is broken and the container cannot easily drain

Figure 16.1
Drainage in gravel beds

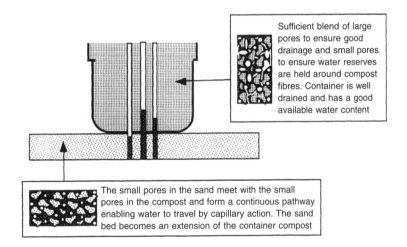

Sufficient blend of large pores to ensure good drainage and small pores to ensure water reserves are held around compost fibres. Container is well drained and has a good available water content

The small pores in the sand meet with the small pores in the compost and form a continuous pathway enabling water to travel by capillary action. The sand bed becomes an extension of the container compost

Figure 16.2
Drainage in sand beds

the sand into the compost. In winter, water drains away through the sand. (See Figure 16.2.)

The aims of a bed are to provide:

● a reservoir of water in summer
● an even water distribution to plants
● drainage from containers in winter.

Exercise 16.1 *Capillary pathways*

Background

A wide range of sand grades may be used for self-watering systems, which when blended together in correct proportions should provide a range of pore sizes to enable water movement by capillary action. Capillary action enables water to move by surface tension from a wetter area to a drier one.

A good sand bed will therefore supply a reservoir of water for plant growth but also enable free drainage of surplus rainfall. Capillary action is stronger with fine sand than coarse sand. Thus, if too many small particles are present water will move upwards through the sand bed rapidly and drainage will be impeded. If too many large particles are present then adequate drainage will be provided but the capillary pathways will not be present, the pores will be unable to pull water into themselves by surface tension and insufficient water will be available to the plants. Figure 16.3 illustrates the capillary action of different particle sizes.

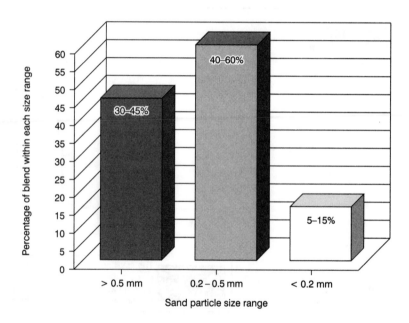

Figure 16.3

Percentage range of each sand grade required to achieve capillary pathways

As a general rule the sand bed should be constructed of a range of pore sizes to enable water to move through a vertical height of 10–23 cm (although the depth of material used in the bed may only be 3–5 cm). The water level in the sand bed is maintained by a header tank (cistern) fitted with a ball valve. This exercise will investigate the varying properties of sand in achieving capillary rise.

Aim

To demonstrate the process of capillary rise.

Apparatus

3 capillary tubes
water
retort stand
coarse sand
medium sand
fine sand
tray

Method

You may find it helpful to draw a labelled diagram of the apparatus.

1 Fill each tube with a different grade of sand.
2 Clamp each tube in a retort stand.
3 Place tubes in a tray of water.
4 Observe the reaction, and record the effectiveness of each material at enabling capillary rise.

Results

1 Measure the water level reached in each tube and record your result in the table provided.
2 Explain the variations in the readings obtained.

Capillary tube	Height of capillary rise (mm)	Explanation of variations
Tube A Coarse sand		
Tube B Medium sand		
Tube C Fine sand		

Conclusions

1 What process is responsible for the rise in the level of the water in the three tubes?
2 Describe the impact of particle sizes on the sand bed's pore structure.
3 Describe the impact of pore size and density on the establishment of capillary pathways.
4 What significance are your findings to commercial horticulture applications?

Exercise 16.2 *Particle size distribution*

Background

Sand beds are a method of self-watering plants eliminating the need for manual operations. They are relatively expensive to construct but yield savings of up to 70 per cent of water, reduce weeds and, through providing a more precise irrigation and drainage regime, substantially increase the quality of container-grown plants.

Some container-grown plants will need to be over-wintered for sale in the following year. In these circumstances root death from waterlogged compost may be as high as 50 per cent and therefore a sand bed will be essential to ensure adequate drainage.

On sites where disease has been a problem a sand bed will be undesirable since many diseases are easily transmitted through water. Large container-grown plants above 15 cm in height will have developed root-balls much less sensitive to water stress and will thrive well with much cheaper overhead irrigation. Similarly plants that are sold within a season will not need over-wintering and therefore the expense of a sand bed construction. In these circumstances a gravel bed will be suitable as a standing material. There are several supplies available from which sand and gravel beds may be constructed. This exercise will investigate some of these materials.

Aim

To investigate the particle size distribution of a variety of standing materials.

Apparatus

> sieve nests (0.06, 0.2, 0.6 and 2.0 mm)
> measuring cylinders (100 and 500 ml)
> 100 ml beakers
> large funnels
> four trays of different standing bed materials, e.g. silver sand, horticultural sand, beach sand, building sand, gravel, labelled A–D

Method

1 Using the sieves provided grade 500 ml of each materials into the following

Fine sand – 0.06–0.2 mm (60–200 μm)
Medium sand – 0.2–0.5 mm (200–600 μm)
Coarse sand – 0.5–2.0 mm (500–2000 μm)
Gravel – 2.0 mm + (<2000 μm)

2 Place each grade into a measuring cylinder and record its volume.
3 Calculate the percentage particle size distribution (PSD) for each material using the following formulae:

$$\% \ PSD = \frac{\text{volume of material within size range}}{\text{starting volume of material sieved}} \times 100$$

4 Retain your sieved materials for use in Exercise 16.3.

Results

Enter your results in the table provided.

Particle sizes	Material A		Material B		Material C		Material D	
	volume (ml)	% PSD	volume (ml)	% PSD	volume (ml)	% PSD	volume (ml)	% PSD
Fine sand								
Medium sand								
Coarse sand								
Gravel								
Total: volume percentage								

Conclusions

1 Define the term 'capillary rise'.
2 Which materials are suitable to construct a gravel bed?
3 Which materials are suitable to construct a sand bed?
4 What is the advantage of a sand bed over placing containers on the ground?
5 What effect will a waterlogged standing area have on root growth?
6 State three circumstances where you might choose to use a gravel bed.
7 State the ideal particle size range for the construction of a sand bed.
8 Explain the principle of sand bed operation.

Exercise 16.3 *Efford sand beds*

Background

Efford experimental station has researched, thoroughly, the range of particle size distribution (PSD) required to achieve a consistent capillary pathway. Its advice is expressed in the background text of Exercises 16.1 and 16.2 and in Table 16.1.

Grade	PSD (%)
Coarse sand	30–45
Medium sand	40–60
Fine sand	5–15

Table 16.1
Efford standards

Aim

To use the materials provided to design an Efford sand bed.

Apparatus

Efford standards (Table 16.1)
materials A–D from Exercise 16.2
seed tray
background from Exercises 16.1 and 16.2

Method

Using your results obtained in Exercise 16.2, design and then construct a model Efford sand bed in the tray provided.

Results

Express your design results in the table provided.

Particle sizes	Source material	Material amount (ml)	% of total material
Fine sand			
Medium sand			
Coarse sand			
Gravel			
Totals:			

Conclusions

1 Which grade of material is the easiest to supply?
2 Which grade of material is most difficult to supply?
3 Explain the advantages of using sand beds in modern container production units.

17 Soil organisms and composting

Background

The soil is teaming with life although, because many organisms cannot be seen with the naked eye, most people are never aware that any life exists. In $1\,m^2$ of woodland soil, there are estimated to be:

- 1000 species of animals
- 10 million nematode worms and protozoa
- 500000 mites and springtails
- 10000 other invertebrates.

In $1\,cm^3$ of soil there are estimated to be:

- 6–10 million bacteria
- 1–2 km of fungal filaments (hyphae).

Functions of soil organisms in composting

Soil organisms are important in horticulture as they are active competitors in decomposing organic matter to produce nutrient rich humus. As soon as a leaf falls to the ground it is subjected to a co-ordinated attack by soil organisms. Bacteria and fungi start this attack and are followed by mites, snails, beatles, millipedes, woodlice and earthworms. (Soil organisms are also pests consuming growing plants, and increasingly are used as agents of biological control preying on one another.) These organisms are referred to as 'decomposers' because their job is to decompose dead organic matter and recycle the nutrients back into the soil for other plants to use.

The role of decomposers is to:

- recycle nutrients
- increase aeration
- stimulate microbial decomposition.

Without decomposers raw materials (nutrients) would run out. Decomposers have the greatest biomass in the system.

Exercise 17.1 *Classification of soil organisms*

Background

The term **flora** refers to vegetative material. **Fauna** refers to animal organisms. The largest animals (macrofauna) feed on leaf/compost litter. The mesofauna (e.g. mites) help circulate nutrients between the litter layer and humus formation. Acting together these organisms mash, digest and oxidize all forms of organic matter including fallen leaves, trunks, dead grass, faeces and defunct bodies, and some species even devour one another.

Soil organisms may be grouped into four categories according to their size and their efficiency at chopping up and decomposing organic matter. This helps to reveal the functional role of decomposers in soil.

Microflora

- bacteria
- fungi.

Microfauna

These are predators of fungal hyphae and bacteria.

- protozoa (amoeba, flagellates, ciliates)
- nematodes (millions per square metre).

Mesofauna

These attack plant litter and the recycled faeces of other soil animals.

- springtails
- mites
- enchytaeid worms
- some fly larvae
- small beetles.

Macrofauna

These are litter feeding invertebrates. They act as mechanical blenders breaking up organic matter and exposing fresh organic surfaces to microbes.

- millipedes
- centipedes
- woodlice
- spiders
- earthworms
- beetles
- slugs
- snails
- harvestmen.

Aim

To gain familiarity with both the the range and classification of soil organisms involved in decomposition and nutrient recycling.

Apparatus

Figure 17.1

Method

1 For each organism listed in Figure 17.1, state whether it is micro-, meso- or macrofauna.
2 Give an example of:

(a) its role in decomposition
(b) an organism which may decompose it.

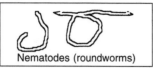 Nematodes (roundworms) — Bacteria, fungi, algae, living plant roots (eelworms). Some are predators of soil animals

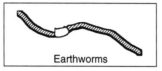 Earthworms — Bacteria, fungi, decaying plant matter

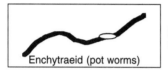 Enchytraeid (pot worms) — Fungi and nematodes

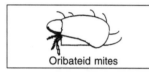 Oribateid mites — Bacteria, fungi, decomposing plant remains

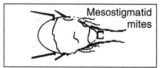 Mesostigmatid mites — Earthworms, potworms, nematodes and mites. Many are parasites of insects used in biocontrol

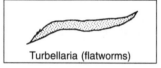 Turbellaria (flatworms) — Microscopic animals in the water film

 Pseudo-scorpions — Predators of mites and springtails

 Spiders — Predators of most insects and small organisms

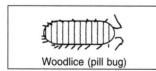 Woodlice (pill bug) — Feed on dead plant and animal matter

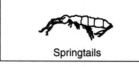 Springtails — Bacteria, fungi and decaying organic matter

 Beetles — Larvae can be very damaging (e.g. wireworms). Consume green leaves of growing pot plants

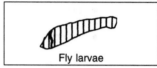 Fly larvae — Feed on decaying or living plants, underground roots and stems (e.g. leatherjackets)

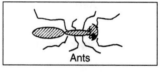 Ants — Consume juvenile soil animals and some plant remains

 Moths and butterflies — Mostly non-feeding pupae, but some larvae feed and live in the soil

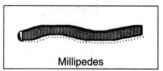 Millipedes — Consume decaying plants and living roots. Important in the mechanical breakdown of humus

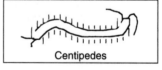 Centipedes — Consume juvenile soil animals

 Slugs and snails — Soil surface feeders of fungi and living plants. Some are carnivorous on earthworms

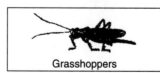 Grasshoppers — Roots, stems and small insect feeders

Figure 17.1
Soil organism and feeding profile

Results

State your results in the table provided.

Soil organism classification			
Micro, meso or macrofauna	Organism	Type of organism to decompose	Type of organism decomposing it
	Snails		
	Mites		
	Springtails		
	Beetles		
	Pseudo-scorpion		
	Millipedes		
	Ants		
	Earthworms		
	Centipedes		
	Spiders		
	Fungal mycelium		
	Slugs		
	Woodlice		
	Larvae		
	Nematodes		
	Bacteria		
	Roundworms		

Conclusions

1 Which classification grouping appears to have the largest population size?
2 Which grouping has the most diverse range of organisms?
3 Distinguish between the terms 'fauna' and 'flora'.

Exercise 17.2 *Biodiversity in a compost heap*

Background

Biodiversity refers to the variety of soil and vegetative organism species present. In a rotting compost heap there will be a tremendous range of organisms present as they are specialized to particular jobs which ensures that all the material is digested and recycled in the form of humus. The macrofauna, for example, will be cutting and blending larger pieces of vegetation which in turn will expose fresh tissue for bacterial and fungal attack. The range of organisms present may be collected in a jar called a pouter (see Chapter 22). This allows you to suck organisms into the jar, but avoids them entering your mouth. Specimens can then be studied either on site or in the laboratory later.

Aim

To investigate and quantify the diversity of decomposing organisms in a compost heap and observe changes with depth.

Apparatus

Figure 17.1	trowel
pouters	sample bags
spades	meter-rules

Method

Cover all wounds to reduce risk from infection.

1 Selected a large, well-rotted compost heap or similar area with decomposing organic matter (e.g. a woodland floor).
2 Visually isolate 1 m^2 of the heap for investigation.
3 Using the pouters, collect and count soil organisms on the surface of the heap.
4 Slowly excavate the heap to depths of 15 cm, 30 cm, 50 cm and 1 m, recording organisms present.

Results

Express your results in the table provided.

Compost heap organism biodiversity results						
	Organisms present					
Depth	Microfauna		Mesofauna		Macrofauna	
	Type	No.	Type	No.	Type	No.
Surface						
15 cm						
30 cm						
50 cm						
1 m						

Conclusions

1 Which group of organisms had the largest population size?
2 Express each classification group as a percentage of the total.
3 Explain how the diversity of organisms changed with depth.
4 What percentage of the total organism population, was each organism:

(a) at each depth?
(b) per cubic meter?

5 Can your results be explained in terms of the type of organic matter present? If so, how?

18 Soil organic matter

Background

Organic matter in the soil is not a well-defined property. It consists of a wide range of compounds forming a biochemical continuum that ranges from single-celled organisms to higher plants and microbes to material of animal origin.

In simple terms organic matter may be considered as all the organic material in soil that can pass through a 2 mm sieve.

Examples of the use of organic matter in horticulture include:

- peat
- farmyard manure (FYM)
- straw incorporation
- mulching.

Organic matter (OM) is substances derived from the decay of leaves and other vegetation, including animal waste and tissues. It may have been broken down into extremely fine humus.

Exercise 18.1 *Soil organic matter determination*

Background

Organic matter can be approximately determined by measuring the organic carbon content of sieved soil. Organic matter itself ranges from 40–60 per cent carbon. When the soil is heated the organic matter burns away (loss on ignition). The mineral fraction is unaffected by the heat, the change in weight therefore approximates to the organic matter content.

Aim

To determine the organic matter status of different soils.

Apparatus

analytical balance	spatula
weigh boats	calculators
metal crucible	goggles
heat resistant mat	dry sieved sample A (lawn soil)
Bunsen burners (2)	dry sieved sample B (orchard
tripod	herbicide strip soil)
forceps	

Method

You may find it helpful to draw a labelled diagram of the apparatus.

1 Collect soil A and soil B from a lawn and an orchard herbicide strip respectively.
2 They must be air dried, ground and sieved to pass through a 2 mm sieve.
3 Zero the balance and weigh the weigh boat (a).
4 Weigh out exactly 50 g of soil in a weigh boat (b).
5 Place soil in the metal crucible.
6 Prepare your furnace by positioning a tripod over a heat resistant mat and locating a Bunsen burner beneath.
7 Using two Bunsen burners heat the tray from below as strongly as possible for 15 minutes. Warning – the tray and soil may become red hot.
8 Allow the tray to cool. Tip the soil into the weigh boat and record the new weight (d).
9 Calculate the OM content using the formula given in the results table (g).

Results

Enter your results in the table provided. The loss in weight represents the organic matter that has been burned off. It should be expressed as a percentage of the weight after heating.

	Determination	Soil A	Soil B
(a)	weight of weigh boat		
(b)	weight of weigh boat and fresh soil		
(c)	weight of fresh soil ((b) – (a))		
(d)	weight of weigh boat and soil after heating		
(e)	weight of soil after heating ((d) – (a))		
(f)	change in weight of soil ((e) – (c))		
(g)	% organic matter content $$\dfrac{\text{Change in weight (f)}}{\text{Weight of soil after heating (e)}} \times 100$$		

Conclusions

1 Explain why the soil sample should be sieved.
2 What sieve size is used?
3 If the soil sample had not been completely dry to start with, how would this have affected your results?
4 What is the reason for heating the soil so strongly and why should this result in a loss in weight?
5 It is possible that 15 minutes' heating is not sufficient to achieve the maximum loss in weight of the soil. What would you have to do to be quite sure that the maximum weight loss had been achieved?
6 Which soil had the greatest organic matter status?
7 Describe how the variations in organic matter status reflect the land use in these two situations.
8 State two benefits from organic matter incorporation into a soil.

Exercise 18.2 *Properties of organic matter*

Background

Organic matter is one of the most important ingredients in a soil. Some of the benefits of organic matter in the soil include:

- high cation exchange capacity (ability for chemical transfers)
- increased water-holding capacity (e.g. 80–90 per cent of weight)
- improved structure and stability
- dark colour, which absorbs sunlight, heating up quickly in the spring and speeding up germination
- improved response to fertilizers.

For these reasons horticulturalists should try to maintain a high level of organic matter in the soil. Organic matter can increase the water-holding capacity of soils. Therefore if organic matter is added to soil, less water will drain away because the volume of water held at field capacity will be artificially higher.

Aim

To demonstrate the increased water-holding capacity of soils with organic matter incorporation.

Apparatus

capillary tubes A and B
sandy soils
retort stand
beaker
measuring cylinder

Method

You may find it useful to draw a diagram of the apparatus.

1 Mix some organic matter (e.g. peat, farmyard manure or straw) with soil A.

2 Fill one capillary tube with soil A, the other with soil B, and fix into retort stands with a beaker beneath each.
3 Pour 100 ml of water into the top of each tube and collect the drainage water in a measuring cylinder.
4 Record the volume of drainage water collected from soils A and B.

Results

Enter your results in the table provided.

Sample	Volume of water added	Volume of drainage water collected	Interpretation
Soil A			
Soil B			

Conclusions

1 Explain which soil has the highest organic matter content by reference to your results.
2 Why is organic matter essential in a fertile soil?
3 Name three types of bulky organic matter commonly added to soil.
4 State two other ways in which organic matter can improve the soil.

Exercise 18.3 *Organic matter and nitrogen*

Background

The breakdown of organic matter liberates nutrients for use by other plants (nutrient cycles). Nitrogen is one such nutrient supplied in this way. Nitrogen is the most important nutrient in plant growth, but is suspected of causing cancer (i.e. is a carcinogen), and is appearing in drinking water in excess of the EEC limit of 50 mg/l.

In this exercise two soils of different organic matter content are compared. The glasshouse soil will have been subjected to intensive cultivation and will be very high in organic matter content from dead roots and vegetative matter which should be clearly visible. The grassland or herbicide strip soil will have less organic matter, particularly if a sample is taken from the subsoil. Nitrogen exists in several different forms in the soil including nitrate, nitrite and ammonium. Nitrate is the most soluble form and the form in which most plants take up nitrogen through their roots. In this exercise all the nitrate that leaches is assumed to have come from the breakdown of organic matter rather than from fertilizer residues.

Aim

To investigate the effect of organic matter on nitrate availability and leaching.

Apparatus

funnels	retort stand
clamps	cotton wool
beakers	soil A (glasshouse soil)
measuring cylinder	soil B (grassland/herbicide strip soil)
distilled water	Merkoquant nitrate test strips

Method

You may find it helpful to draw a labelled diagram of the apparatus.

1 Place a small cotton wool ball in the neck of a funnel, held in the retort stand.
2 Repeat step 1 with another funnel.
3 Add soil A, and soil B to different funnels .
4 Place a beaker beneath each funnel.
5 To soils A and B, slowly add 100 ml additions of distilled water until the first drops of leachate (drainage water) appear.
6 Place a nitrate test strip in the leachate for 1 second, then, after a further 60 seconds (after which time full colour development will be complete) compare the colours against the standard colour chart provided with the test strips to ascertain the nitrate content.
7 Pour away the remaining leachate and rinse the beaker with distilled water before replacing.
8 Continue to add further 25 ml additions, until an additional 200 ml has been added, discarding the leachate between additions.
9 Measure the nitrate content again as in step 6.

Results

Record your results in the table provided.

Soil Type	Nitrate content (Mg/l)		% change
	Leachate 1	Leachate 2	
Soil A (glasshouse)			
Soil B (grassland/herbicide strip)			

Conclusions

1 Which soil had the highest nitrate content?
2 How can your results be explained by reference to the organic matter content of each soil?
3 Which soil is more susceptible to nitrate leaching?
4 Comment on how organic matter content may be affected by earthworms and cultivations.

19 Plant nutrition

Background

As with all living things, plants need food (nutrition) for their growth and development. Plants live, grow and reproduce by taking up water and mineral substances from the soil, together with carbon dioxide from the atmosphere and energy from the sun, to form plant tissue.

A large number of elements are found in plant tissue. Of these, sixteen have been found to be indispensable without which the plant would die. These elements are referred to as nutrients.

Nutrients from the atmosphere or soil water

Carbon (C)
Hydrogen (H)
Oxygen (O)

Nutrients from the soil/growth medium

Nitrogen (N)
Phosphorus (P)
Potassium (K)
Calcium (Ca)
Magnesium (Mg)
Sulphur (S)
Iron (Fe)
Zinc (Zn)
Manganese (Mn)
Copper (Cu)
Boron (Bn)
Molybdenum (Mo)
Chlorine (Cl)

Plants have also been found to benefit from supplies of cobalt (Co), sodium (Na) and silicon (Si), but these are not considered 'essential' nutrients. That is, without them, the plant will not die.

The atmospheric gases (C, H and O) are essential for photosynthesis. They are also used in making plant carbohydrates and proteins.

Of the soil-derived nutrients, nitrogen, phosphorus and potassium are used in the largest quantities, followed by magnesium, calcium and sulphur. This may be as little as 1 kg/ha up to several 100 kg/ha, and they are therefore referred to as **major nutrients** (or macronutrients).

The remaining nutrients (Fe, Cu, B, Mo, Cl, Mn and Zn) are required in smaller quantities, and in large amounts are toxic to plants. These are referred to as **trace elements** (or micronutrients).

The soil or growth media must contain all the essential nutrients in sufficient quantity and in balanced proportions. These nutrients must also be present in an available form before plants can use them. An inadequacy of any one of these elements will inhibit plants from growing to their full potential.

Exercise 19.1 *Plant nutrition*

Aim

To develop knowledge of the ingredients of plant nutrition.

Apparatus

background text section

Method

Read the following questions and phrase suitable answers.

Results

1 What is a plant nutrient?
2 How many plant nutrients are there?
3 Explain what a major nutrient is.
4 List the major elements.
5 Explain what a trace element is.
6 List the trace elements.
7 What effect will an over-application of trace elements have on plant growth?
8 What other terms are used to describe:

(a) major elements?
(b) trace elements?

9 Which nutrients are mostly supplied as gases?

Conclusions

1 What nutrition is required to support plant growth?
2 Explain what effect an absence of a plant nutrient will have on plants?
3 Explain the impact of an inadequate supply of any nutrient on growth.
4 Why are sodium, silicon and cobalt not 'technically' plant nutrients?

Exercise 19.2 *Tomato plant deficiency symptoms*

Background

The ability to identify nutrient disorders correctly is extremely helpful in diagnosing plant health problems. The problems will mostly arise in pot plants where the compost may have been mixed on the nursery or where supplementary overhead irrigation also supplies some nutrition. Typically, a fertilizer such as phosphorus will have been omitted from the compost mix and the plants will grow with purple leaf edges or other symptom of the deficiency. In soils, over-cropping and soil texture variations may induce a deficiency. In either case, the problem must be identified and corrective action taken.

In this exercise tomato seeds are sown in an inert material (such as rockwool or perlite) and watered using a solution which lacks one nutrient. Tomatoes are a particularly good plant to grow since they demonstrate very easily the deficiency symptom under study.

The plant grown with a complete range of nutrients should demonstrate healthy growth and colouring. Tap water contains some nutrients but in very low quantities. Distilled water is the purest form of water available and will contain no nutrients at all. The tomatoes watered with this will therefore have to survive just on the energy contained in the germinating seed. The remaining plants will show varying degrees of leaf colour, growth, root development and stem strength which should all be observed.

Aim

To gain increased competence at the visual diagnosis of nutrient deficiency symptoms.

Apparatus

> tomato seeds
> rockwool cubes or perlite growing medium
> tomato plant liquid feed solutions lacking: nitrogen, phosphorus, potassium, magnesium (e.g. The 'Long Ashton Water Culture Kit', from Griffen and George (see List of Suppliers), provides premixed chemicals that are dissolved in water)
> tomato plant liquid feed with tap water and distilled water
> tomato plant feed with complete range of nutrients

Method

You may find it useful to draw a diagram of the deficiency symptoms shown in the plants.

1 Sow half a dozen seeds into each of seven rockwool cubes (or perlite) on the same day (enough for each of the above treatments).

2 Water each seed and subsequently germinating plant with one of the above solutions.

3 If the cube becomes crowded, thin out the plants leaving those that demonstrate good symptoms of the deficiency under study.

4 After several weeks, compare the growth profiles of the tomato plant receiving the complete feed with those lacking N, P, K, Mg, and those fed with tap water and distilled water, recording the visual symptoms of the deficiency.

Results

Record your results in the table provided.

Type of feed	Description of condition and symptoms of deficiency
Complete nutrient range	
Distilled water	
Tap water	
Lacking N	
Lacking P	
Lacking K	
Lacking Mg	

Conclusions

1 Why was distilled water used in this experiment?

2 Explain why you would not rely on tap water as a nutrient supply, especially when feeding cut flowers in a vase.

3 Which nutrient deficiency is most easily recognized?

4 Assuming that the deficiency symptoms that you observed are representative of the typical symptoms likely to be found in a range of plants, identify the deficiencies present in each of the following situations:

(a) A chrysanthemum suffering from stunted growth and light green leaves.

(b) Poinsettia containing yellow leaf edges and pale green/yellow blotches between leaf veins. The leaves are small and brittle, turning upwards at their edges and suffer premature leaf drop.

(c) Alstroemeria having a scorched yellow look to the edges of the older leaves, stalks are weak and the plants collapse (go floppy) easily.

(d) Solanums with leaf edges containing a thin margin of purple colouring.

Exercise 19.3 *Function of plant nutrients*

Aim

To recognize the physiological importance of nutrients to plant development.

Apparatus

list of nutrients: C, H, O, N, P, K, Mg, Ca, S, trace elements
tomato plant specimens from Exercise 19.2
Table 19.1

Method

Observing the tomato plants and referring to the literature match the following list of nutrient functions with the nutrient responsible.

Results

Which nutrient:

(a) increases the vegetative growth?
(b) is a component of carbohydrate and proteins and is essential in photosynthesis (two nutrients)?
(c) is a component of carbohydrate and proteins and is essential in photosynthesis and respiration?
(d) controls osmosis and resistance to disease?
(e) is a constituent of chlorophyll?
(f) improves fruit storage life and is a constituent of cell walls?
(g) is a component of some amino acids?
(h) has many functions mainly connected with photosynthesis, nitrogen assimilation and protein formation?
(i) is responsible for root development in young plants?

Conclusions

1 Explain the difference between a major and a minor nutrient?
2 How are primary major nutrients distinguished from secondary major nutrients?
3 Which nutrients most commonly need to be applied as fertilizers?

Exercise 19.4 *Major nutrient roles*

Background

The primary major nutrients are the nutrients that most often suffer deficiency and will need to be applied as fertilizers. It can be useful to remember their roles in the plant by reference to a graphical representation of a plant with the function of the nutrient closely associated with its position on the plant diagram.

Aim

To consolidate learning of the role played by nutrients in plant physiology.

Table 19.1
The role of plant nutrients and their deficiency symptoms

Nutrient	Role in plant	Susceptible soils	Deficiency symptoms
Carbon and hydrogen	• component of all carbohydrates and proteins • essential in photosynthesis	Supplied as atmospheric gases and in the soil solution	Deficiencies unknown
Oxygen	• component of carbohydrates and proteins • essential in photosynthesis and respiration		
Nitrogen	• Essential for vegetative growth • Protein and chlorophyll formation	Light textured soils lacking organic matter	• Stunted growth • Light green to pale yellow leaves starting at the leaf tip, followed by death of the older leaves if the deficiency is great • In severe cases the leaves turn yellowish-red along their veins and die off quickly • Flowering is both delayed and reduced
Phosphorus	• Essential for root growth during the early stages of plant life • Helps plant to flower and fruit productively • Enables meristematic growth • Component of amino acids and chlorophyll • Necessary for cell division	Occasionally found on heavy textured soils and peat composts	• Bronze to red/purple leaf colour • Shoots are short and thin • Older leaves die off rapidly • Flower and seed production are inhibited • Restricted root development • Delayed maturity

Potassium	Essential for many physiological reactions within the cell:
	  • Control of osmosis
	  • Resistance to drought, disease and frost
	  • Influences the uptake of other nutrients, e.g. Mg
	  • Flower and fruit formation
	  • Improves the quality of seed, fruit and vegetables
		Occasionally found on light textured soils	• Stunted growth
			• Scorched look to the edges of older leaves (chlorosis) gradually progressing inwards
			• Stalks are weak and plants collapse easily
			• Shrivelled seeds or fruit
			• Brown spots sometimes develop on leaves

Nitrogen, phosphorus and potassium are the three most important macro-elements and are referred to as the 'primary major nutrients'

	Functions	Soil occurrence	Deficiency symptoms
Magnesium	• Essential for chlorophyll formation • Influences phosphorus mobility • Influences potassium uptake by the roots • Helps the movement of sugars within the plant	• Often found on light textured, sandy or peaty soils, especially in high rainfall areas where there is an excess of calcium or potassium	• Leaves develop yellow margins and pale green/yellow blotches between veins (interveinal chlorosis) • Blotches become yellower and eventually turn brown • In final stages leaves are small and brittle, edges turn upwards • In vegetables, plants are coloured with a marbling of yellow with tints of orange, red and purple • Stems are weak and prone to fungal attack • Premature leaf drop
Calcium	• Essential for development of growth tissue (tips) • Constituent of cell walls • Essential for cell divisions, especially in roots • Maintenance of chromosome structure • Acts as a detoxifying agent by neutralizing organic acids in plants • Produces firmer fruit • Prevents bitter pit in apples and cork spot in pears • Improves storage life	• Normally, if the pH of the soil is satisfactory, then so is calcium content. Deficiencies are rare but can be found on light textured, acid or peaty soils	• Deficiencies are not often seen, partly because secondary effects associated with high acidity tend to limit growth • Chlorosis of the young foliage, and white coloration to edges • Death of the terminal bud • Growing point may shrivel up and die • Grey mould (botrytis) infection to growing point • Distorted leaves with the tip hooked back • Death of root tips and damaged root system (appearing rotted) • Buds and blossoms shed prematurely • Stem structure weakened
Sulphur	• Component of amino acids, proteins and oils • Involved with activities of some vitamins • Aids the stabilization of protein structure	• Sulphur is a common atmospheric pollutant and therefore rarely added as a fertilizer	• Uniform yellowing of new and young foliage

These are the secondary major nutrients, but are not required in the same quantities as N, P or K

	Functions	Soil occurrence	Deficiency symptoms
Trace elements	• Many functions mainly associated with photosynthesis, nitrogen assimilation or protein formation	• Rare in soils but more common in composts	• Various symptoms including brittle leaves, chlorosis, leaf curling, poor fruiting and distorted leaf colours

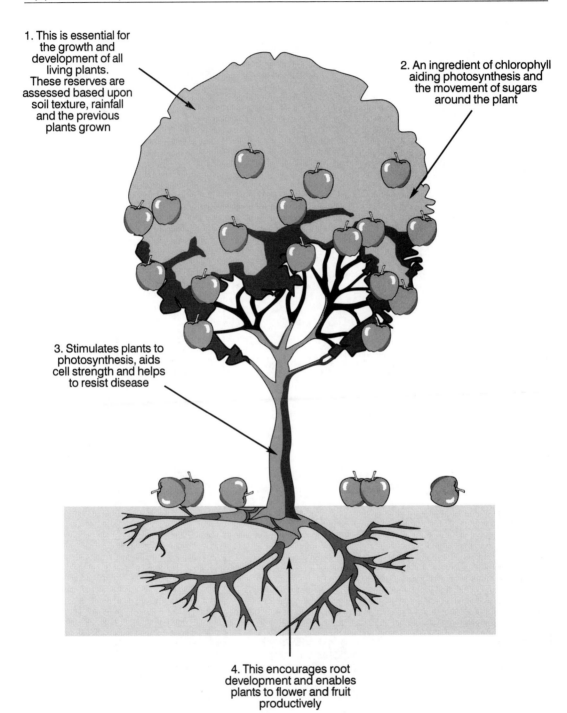

1. This is essential for the growth and development of all living plants. These reserves are assessed based upon soil texture, rainfall and the previous plants grown

2. An ingredient of chlorophyll aiding photosynthesis and the movement of sugars around the plant

3. Stimulates plants to photosynthesis, aids cell strength and helps to resist disease

4. This encourages root development and enables plants to flower and fruit productively

Figure 19.1
Plant nutrient roles

Apparatus

Figure 19.1
Table 19.1

Method

Observe the plant illustrated in Figure 19.1 and decide which nutrients are being described in the text.

Results

Enter your answers on the dotted lines provided.

1 . . . .

2 . . . .

3 . . . .

4 . . . .

Conclusion

Define the terms 'major element' and 'minor element'.

Exercise 19.5 *Nitrate fertilizers*

Background

Nitrogen is probably the most important nutrient. The deficiency is fairly common, especially on light textured soils lacking organic matter. It is used in protein and chlorophyll formation and is very effective at stimulating vegetative growth. Most plants prefer to take up nitrogen in the form of nitrate (NO_3), but Ericaceae species prefer it as ammonium (NH_4). Nitrite (NO_2), is toxic to most plants. Nitrogen levels are extremely difficult to measure because:

(a) it is highly soluble and will be leached out quickly, so readings for fertilizer applications are useless
(b) nitrogen is continually being mineralized (recycled) from soil organic matter.

Nitrogen should therefore be applied as little as possible as often as possible. For the purpose of fertilizer applications, therefore, nitrogen levels are 'estimated' based upon:

(a) previous plants grown
(b) summer rainfall
(c) soil texture.

Fertilizers may be used in one of four different ways to supplement soil reserves: base dressing, top dressing, foliar feed or liquid feed (see fertilizer exercises). Some fertilizers are more suited to specific applications than others due to their differing rates of nutrient release characteristics. In this exercise the release rate of nitrogen from organic fertilizer is compared with that of an inorganic fertilizer.

Aim

To appreciate differences in the release rate of nitrogen from fertilizers.

Apparatus

funnels	nitrate test strips
retort stand	beakers
clamps	balance
cotton wool	sandy soil
measuring cylinder	ammonium nitrate
distilled water	hoof and horn

Method

1 Weigh out 1 g of each fertilizer and mix with 100 ml of distilled water.
2 Record the nitrate content of the solution using the test strips.
3 Record your result in the table provided.
4 Place a small piece of cotton wool in a funnel and position in a retort stand above a beaker.
5 Fill the funnel with a small amount of sandy soil.
6 Repeat steps 4 and 5 to make a second set.
7 Weigh out 5 g of inorganic fertilizer (ammonium nitrate) and broadcast on the top of one soil.
8 Weigh out 5 g of organic fertilizer (hoof and horn) and broadcast on the top of the other soil.
9 Slowly add 25 ml of distilled water, in increments, until the first leachate is collected.
10 Using the Merkoquant nitrate test strip, record the nitrate content of the leachate.
11 Continue to add further 25 ml additions, until an additional 200 ml has been added, discarding the leachate between additions.
12 Record the resulting leachate nitrate content.

Results

Record your results in the tables provided.

Fertilizer solution	Nitrate content (Mg/l)
1% w/v Ammonium nitrate	
1% w/v Hoof and horn	

| Soil treatment | Nitrate content (Mg/l) | | % Change |
	Leachate 1	Leachate 2	
Soil + ammonium nitrate			
Soil + hoof and horn			

Conclusions

1 Which fertilizer contains the highest nutrient content?
2 Which fertilizer is more soluble?
3 Which fertilizer treatment would result in more nitrate leaching?
4 State an application use for the two fertilizers with reference to the release rate of nitrogen, giving reasons for your answers (e.g. top dressing, base dressing, etc.):

(a) Sulphate of ammonia
(b) Hoof and horn

5 How should nitrogen be applied as a fertilizer?
6 Explain the role of nitrogen in plant growth and development.
7 Describe the typical symptoms of nitrogen deficiency in a named plant.

20 Fertilizers

Background

Fertilizers are a method of applying nutrients to the soil to enhance plant growth. The amount of fertilizer added is based upon existing soil reserves of available nutrients and the intended plants to be grown. The quantity added should be sufficient to meet plant growth without wastage (e.g. due to leaching). This requirement affects the type of fertilizer used (e.g. granular, prill, slow release) and the timing of applications (e.g. nitrogen, as little as possible, as often as possible).

Application of fertilizers

Fertilizers can be applied in one of four different ways:

Base dressing – mixed into growth media, usually before planting

Top dressing – broadcast onto the soil/growth medium surface

Foliar feed – normally to correct deficiencies is sprayed onto leaves

Liquid feed – in hydroponics production systems is normally supplied direct to the roots.

Generally, fertilizers are used to supply nitrogen (N), phosphorus (P), potassium (K) or combinations of these. In addition magnesium (Mg) is often needed, together with trace elements. A range of fertilizers used to supply N, P, K and Mg is presented in Table 20.1.

Nitrogen

Most plants absorb nitrogen as nitrate which is extremely soluble making leaching a problem. The quantities of leached nitrates appearing in water supplies has often exceeded the EU limit of 50 mg/l. Nitrogen should be added as little as possible, as often as possible to avoid this. Many people now add nitrogen in a slow release form. Additional nitrogen fertilizers include nitrate of soda (16 per cent N), nitrochalk (21 per cent N), urea (45 per cent N), urea formaldehyde (40 per cent N), Gold N (36 per cent), Nitroform (38 per cent N) and Nitrochalk (26 per cent N), among others.

Phosphorous

Phosphorus is extremely insoluble. At pH 6.5 it is 70 per cent insoluble, and in acid and alkaline conditions 100 per cent insoluble. Phosphorus is bound tightly to clay particles but can be washed out of growth media easily. Fungi associations help plants to take up P. Where possible water-soluble fertilizers should be used. Manufacturers will always state on the bag what percentage of the mix is water and acid soluble. Phosphorus content is stated as phosphorus pentoxide (P_2O_5). Additional phosphorus fertilizers include rock phosphate (25–40 per cent), basic slag (12–18 per cent), di-ammonium phosphate (54 per cent) and blood, fish and bone (3.5–8–0.5 variable).

Potassium

Potassium is generally readily soluble in water, but is not easily leached. Luxury supply of potassium should be avoided. Potassium is associated with the health and efficient functioning of plants and is generally applied in proportion to nitrogen. The potassium content of fertilizers is declared potash (K_2O), otherwise called potassium oxide. Additional potassium fertilizers include Chilean potash nitrate (10–15 per cent), Kainit (14–30 per cent), and farmyard manure (40 per cent).

Magnesium

Magnesium is required in about 10 per cent of the plant's N and K, requirements, roughly equal to P, and 100 times greater than trace elements. Most soils have adequate reserves of Mg. Magnesium content is stated as % Mg. Additional Mg fertilizers

Table 20.1
Nutrient content of some common straight fertilizers

Fertilizer name	Nutrient content	Solubility	Release rate	Form	Characteristics	Uses
Nitrogen						
Ammonium nitrate 33.5–34.5% N. e.g. Nitram. This is the commonest N fertilizer in the world.	33.5/34.5-0-0	50% is immediately soluble and 50% slowly soluble	Very fast	Inorganic	Tends to make soils acid so needs extra lime. Useful early in season for overwintered crops	Top dressing Liquid feed Base dressing
Sulphate of ammonia 21% N. Most favoured amateur N fertilizer. Cheap	21-0-0	Soluble	Fast	Inorganic	Has a greater acidifying action than all other nitrogen fertilizers and is less efficient than nitrate forms	Top dressing Liquid feed Base dressing
Calcium nitrate 15% N. Does not acidify the soil. More expensive than other N sources	15.5-0-0	Soluble	Fast	Inorganic	Mainly used on horticultural plants where acid soils are not desired	Top dressing Liquid feed Base dressing
Dried blood 10–13% N. Quick-acting organic fertilizer	12-13-0	Partly soluble	Fairly to very fast	Organic	Especially suited for liquid feeding	Liquid feed Top dressing
Hoof and Horn 13% N. Ground-up hoofs and horns of cattle	13-14-0	Insoluble slow release	Slow to fairly fast	Organic	Rate of release depends upon size of particles	Base dressing Top dressing
Phosphorus						
Mono ammonium phosphate Best known of the water-soluble forms	12-25-0	Soluble	Fast	Inorganic	Made from rock phosphate, ammonia and phosphoric acid	Liquid feed Top dressing
Superphosphate Also known as single superphosphate	0-18/21-0	Soluble	Fast	Inorganic	Contains calcium sulphate which is acidic and can raise conductivity	Top dressing Base dressing

Material	N-P-K					Application
Triple superphosphate Another water-soluble straight P source	0-47-0	Soluble	Fast	Inorganic	Phosphorus rock mixed with acid. Contains no impurities. Neither form of superphosphate is suitable for liquid feeding	Top dressing Base dressing
Bone meal (should be sterilized)	0-22-0	Slowly soluble	Slow	Organic	Expensive form of P, that contains no advantages over the inorganic forms. Also contains a little N	Base dressing
Steamed bone meal/flour (should be sterilized)	0-28-0	Slowly soluble	Slow	Organic	Both bone meals are in little demand, except for market garden and amateur use	Base dressing

Potassium

Material	N-P-K					Application
Sulphate of potash	0-0-50	Soluble	Fast	Inorganic	Manufactured to form a dense powder that does not cake easily	Base dressing Top dressing Liquid feed
Potassium nitrate	13-0-44	Fast	Soluble	Inorganic	Also known as saltpetre. Expensive. Mainly used as a liquid feed	Liquid feed Top dressing
Muriate of potash (potassium chloride)	0-0-60	Soluble	Fast	Inorganic	Small dry crystals that do not cake easily. Cheap, often used on fruit, but chloride content can raise conductivity and damage soft fruits	Foliar feed Base dressing
Rock potash	0-0-50	Slow	Slowly soluble	Organic	Difficult to obtain the natural mineral and rarely used. Needs to be ground	Base dressing

Magnesium

Material	N-P-K					Application
Epsom salts (magnesium sulphate)	0-0-0-10	Soluble	Fast	Inorganic	Useful for foliar spray or liquid drench. This is an expensive form of Mg	Foliar feed Liquid feed Top dressing Base dressing
Kieserite (another form of magnesium sulphate)	0-0-0-17	Slowly soluble	Very slow	Inorganic	Mg becomes available to plants in the first season, and then for the following two or three seasons	Base dressing
Magnesian limestone (dolamitic limestone)	0-0-0-8/12	Slowly soluble	Slow	Inorganic	Used when magnesium levels are low and acidity needs to be corrected	Base dressing

are rare. Farmyard manure has been found to contain 1 per cent, basic slag (0.5 per cent). In the UK, rainfall provides 2 kg/ha for each 100 mm of rain.

Fertiliser bag content

Fertilizers list N-P-K-Mg as a percentage of the weight of the fertilizer, e.g. Osmocote (18-11-10), Ficote 140 (16-10-10). Thus, Osmocote is 18 per cent N, 11 per cent P and 10 per cent K; similarly, Ficote 140 is 16 per cent N, 10 per cent P and 10 per cent K. The remaining material is impurities and a carrier material such as sand, used to aid spreading the fertilizer.

Exercise 20.1 *Fertilizer nutrient content*

Aim

To increase knowledge of general N, P, K and Mg fertilizer frequently used in horticulture.

Apparatus

background text
Table 20.1

Method

Answer the list of questions provided.

Results

1 State the percentage nitrogen content of ammonium nitrate.
2 List an organic nitrogen fertilizer suitable for use as a base dressing prior to planting shrubs.
3 Explain the difficulties in applying phosphorus as a fertilizer.
4 State the percentage phosphorus content and the uses of triple superphosphate.
5 Sulphate of potash is a common garden centre retailed fertilizer. State its nutrient content.
6 Explain why rock potash is unsuitable for use as a foliar feed.
7 Epsom salts is a very common form of magnesium fertilizer. What is its chemical name and nutrient content?
8 Select a slow release magnesium fertilizer suitable for use as a base dressing.

Conclusions

1 Explain the difference between a 'base dressing' and a 'top dressing'.
2 Explain the difference between a 'liquid feed' and a 'foliar feed'.
3 A rose fertilizer has a nutrient content of 5-6-12. Explain what this means.

Exercise 20.2 *Structure of fertilizers*

Background

There are three categories of solid fertilizer available to horticulturalists. These are:

Powdered fertilizer in the form of powder or dust (e.g. hoof and horn).

Granulated fertilizer which comprises non-spherical grains from 1 to 4 mm in diameter. They tend to be rough in texture, and irregular in both size and shape. Granular materials may be straight N, P, K, or any proportion of each. Straight N fertilizer is particularly common to this group. The materials are hard and have a higher density than powdered, therefore making them easier to spread.

Prilled fertilizer comprises spherical balls of uniform size (approximately 2 mm in diameter), but can vary from less than 1 mm up to 3.5 mm. They are round, smooth and very free flowing. They can be coated to make them harder and less susceptible to moisture absorption. Straight nitrogen is the most common prilled fertilizer, either as ammonium nitrate or as urea.

Aim

To identify correctly the structure of fertilizers.

Apparatus

weigh boats	kieserite
ammonium nitrate	ficote
steamed bone flower	Table 20.1
copper sulphate	

Method

1 Decant a small amount of each fertilizer into a weigh boat.
2 Observe the size, shape and texture of the fertilizers and decide whether they are powdered, granulated or prilled.
3 Reading from Table 20.1 or other sources, state the type and percentage nutrient supplied.

Results

Enter your results in the table provided.

Fertilizer	Structure	Nutrient supplied	% of nutrient in fertilizer
Ammonium nitrate			
Steamed bone flower			
Copper sulphate			
Kieserite			
Ficote			

Conclusions

1 Explain, based upon your observations, which fertilizers may be used for:

(a) base dressing?
(b) top dressing?

2 State two other methods of applying fertilizers?

Exercise 20.3 *Fertilizer spreading*

Background

The structure of fertilizers is important, largely due to the ease of spreading. This in turn is influenced by:

(a) particle size (prills, granules, etc.)
(b) density
(c) hardness of material.

These properties combined can add up to the difference between throwing a ping-pong ball or a golf ball. Materials with larger particle sizes and higher densities will be better for spreading.

Material hardness

This test can also be used to assess both the age and solubility of fertilizers. Older fertilizers crumble more easily as do more soluble ones. Try to crush some fertilizer with the back of your finger nail. If the fertilizer:

- crumbles, the product is poor
- crumbles with difficulty, the product is good
- is impossible to crush, the product is very good.

Aim

To assess the age, solubility and ease of spreading of fertilizers.

Apparatus

weigh boats	copper sulphate
ammonium nitrate	kieserite
steamed bone flower	ficote

Method

Grade the fertilizers using the material hardness test.

Results

Enter your results in the table provided.

Fertilizer	Material hardness	Interpretation
Ammonium nitrate		
Steamed bone flower		
Copper sulphate		
Kieserite		
Ficote		

Conclusions

1 Which materials would be the easiest to spread?
2 Which materials are the oldest stock?
3 Which materials are probably the most soluble?
4 List six major nutrients required by plants for healthy growth.
5 Name four fertilizers commonly used as top dressings.

Exercise 20.4 *Types of fertilizer*

Background

There are three basic types of fertilizer formulation available:

1 **Straight** – These contain only one of the major plant nutrients, e.g. ammonium nitrate
2 **Compound** – These fertilizers contain two or more nutrients bonded together, e.g. Growmore 7-7-7.
3 **Mixed/blended** – These are produced by mixing two or more straight fertilizers together, e.g. John Innes Base Fertilizer; not a common method.

Aim

To identify types of fertilizers.

Apparatus

weigh boat
ammonium nitrate
urea
potassium nitrate

ammonia sulphate
sulphate of potash
hoof and horn
Table 20.1

Method

1 Decant a small quantity of each fertilizer into a weigh boat.
2 Examine each material with its packaging, and decide whether each fertilizers is straight, compound, or mixed/blended.

Results

Enter your results in the table provided.

Fertilizer	Straight, compound or blended	Nutrient(s)	% supplied
Ammonium nitrate			
Urea			
Potassium nitrate			
Ammonium sulphate			
Sulphate of potash			
Hoof and horn			

Conclusions

1 Which fertilizer might you use as a top dressing in summer for lawns?
2 Which fertilizer might you use as a base dressing for a shrubbery in spring?
3 State what is meant by:

(a) straight fertilizer?
(b) compound fertilizer?

4 State a fertilizer that may be used as a foliar feed to correct nitrogen deficiency.

Exercise 20.5 *Speed and mode of nutrient release*

Background

Fertilizers may also be classified according to how fast or slow they release their nutrients into the soil or growth medium.

Fast release

With all these fertilizers the nutrients are available quickly and can be taken up by the plant. This group includes those that are readily soluble. Often horticulturalists rank these according to the speed of reaction of the plant to the fertilizer.

Slow release/controlled release

These may be tablets, granules, powders or even stick fertilizers whose nutrition is not immediately soluble. They release their nutrients slowly over an extended period of time based upon their rate of decomposition/ mineralization. Three such groups can be identified:

Organic fertilizers. Nutrients become available by the slow breakdown and mineralization resulting from the actions of micro-organisms, e.g. bone meal.

Slowly soluble fertilizers. These break down slowly as water absorbs and bacteria degrade into them over longer periods of time than those fertilizers that are immediately soluble, e.g. urea formaldehyde.

Controlled release (C-R-F). These are resin or polymer coated fertilizers. Their skins gradually decompose to release nutrients by the action of water and temperature.

New products are appearing regularly. The most widely used C-R-F are Osmocote and Ficote. Rates of release vary from eight weeks to eighteen months. Generally, however, they may be divided into two:

(a) single season (8–9 months), e.g. Ficote 70 (16-10-10)
(b) extended season (12–18 months), e.g. Ficote 140 (16-10-10).

When the resin coats absorb water their contents are released slowly, through submicroscopic pores. The major factor controlling the rate of flow of nutrients through the skin is temperature. As the temperature increases so does the porosity of the coating. In this way the release of nutrients matches the crop demand. To prevent excess build-up of nutrients in the growth media, the porosity decreases, cutting off the supply of nutrients if the temperature rises too high.

Aim

To recognize modes of fertilizer release.

Apparatus

weigh boats	mono ammonium phosphate
osmocote	beaker
ficote	water
ammonium nitrate hoof and horn	spatula
dried blood	Table 20.1
potassium nitrate	

Method

1 Decant some of each fertilizer into a weigh boat.
2 Conduct a material hardness test (see Exercise 20.3) and note the fertilizer response when mixed in the beaker with a little water.
3 Decide whether each fertilizer is fast, organic, slowly soluble or controlled release.

Results

Enter your observations in the table provided.

Fertilizer	Speed of nutrient release (fast, organic, slow or controlled)
Osmocote	
Ficote	
Ammonium nitrate	
Hoof and horn	
Dried blood	
Potassium nitrate	
Mono ammonium phosphate	

Conclusions

1 Based upon your observations, state which of these fertilizers may be used for:

(a) base dressing
(b) top dressing
(c) foliar feed
(d) liquid feed.

2 Describe the role of controlled release fertilizers and fast release fertilizers in horticulture.
3 Explain why hoof and horn fertilizer may be ineffective when used in spring as a nitrogen top dressing.
4 Name a fertilizer suitable for each of the following uses:

(a) an acidifying nitrogen fertilizer for turf or field use
(b) a fertilizer that will supply phosphate for liquid feeds
(c) a fertilizer for liquid feeding that is also a liming material
(d) a magnesium fertilizer that is also a liming material.

5 State two examples of slow release fertilizers suitable for use with container-grown plants.

Exercise 20.6 *Fertilizer recommendations and calculations*

Background

Fertilizer recommendations are based upon soil analysis and the requirements of the plant to be grown, making an allowance for nutrient residues left from the previous plants grown. All fertilizer recommendations, such as from analysis laboratories, Agricultural Development Advisory Service (ADAS), reference books, amateur journals (e.g. *Gardening Which?*) and in manufacturers' data sheets, are reported in kg/ha, e.g.:

N = 220 kg/ha (N)
P = 250 kg/ha (P_2O_5)
K = 300 kg/ha (K_2O).

This may be applied either as straight fertilizers or by choosing a compound fertilizer to give as near as possible the correct amount. In this case the priority is to get the nitrogen supply right. Slight variations in the rate of phosphorus and potassium will have less effect.

Aim

The end user will have to interpret this recommendation according to the fertilizers they have in stock. You must therefore be able to translate the recommendation into its equivalent for the fertilizer that you have available. This can be achieved by using the following formula:

$$\text{Amount of fertilizer required in kg/ha} = \frac{\text{Nutritional requirement advised (in kg/ha)}}{\text{\% nutrient concentration in alternative fertilizer}} \times 100$$

Method

Work through the two example calculations below, and then answer the questions. You need not answer every question; when you have confidently mastered this skill, proceed to Exercise 20.7.

Example 1

ADAS recommend applying 36 kg/ha phosphorus pentoxide (P_2O_5). The only fertilizer available is single superphosphate (18 per cent P_2O_5).
Using the above formula:

$$\text{Single superphosphate required in kg/ha} = \frac{36 \text{ kg/ha}}{18\%} \times 100$$
$$= 2 \times 100 = 200 \text{ kg/ha}$$

To convert kg/ha to g/m^2, divide by 10, e.g.

200 kg/ha divided by 10 = 20 g/m^2 of single superphosphate.

Example 2

ADAS recommend 36 kg/ha phosphorus pentoxide and you have triple superphosphate (TSP) (46 per cent P_2O_5) available:

$$\text{Amount of TSP required } = \frac{36\,\text{kg/ha}}{46\%} \times 100$$

$$= 0.78 \times 100$$

$$= 78\,\text{kg/ha triple superphosphate}$$

$$= \frac{78}{10} = 7.8\,\text{g/m}^2$$

Results

1 Calculate the amount of mono ammonium phosphate (25 per cent P_2O_5) required to provide a top dressing recommended in *Gardening Which?*, of 36 kg/ha phosphorus pentoxide.

2 Calculate the amount (in kg/ha and g/m^2), that would be required to provide a base dressing of potash (K_2O) of 150 kg/ha, using muriate of potash (60 per cent K_2O).

3 A handbook for lawns indicates that the lawns require a top dressing of 100 kg/ha nitrogen (N). How much ammonium nitrate (33.5 per cent N) should you use?

4 As a result of a soil analysis report, the *Elaeagnus pungens* stock plant beds require a top dressing of 50 kg/ha N. How much sulphate of ammonia (21 per cent N) should you use?

5 Analysis results show that the soil in the all-year-round chrysanthemum house is becoming unacceptably acidic, due to the previous fertilizer regime. The nitrogen recommendation is for 200 kg/ha N. You have checked the stores and have both sulphate of ammonia (21 per cent N) and calcium nitrate (15 per cent N) available. Which one should you use? Why? And at what application rate?

6 The lettuce plants require a top dressing of 150 kg/ha N. The manager has decided to grow them organically because she believes that this will yield a premium price. How much dried blood (12 per cent N) should be used?

7 A local grower have decided to establish an organically grown Pick-Your-Own strawberry unit. How much hoof and horn (13 per cent N) should he use to meet the recommended base dressing rate of 75 kg/ha?

8 How much single superphosphate (18 per cent P_2O_5) is required to complete a turf seed bed preparation rate specified at 75 kg/ha phosphorus pentoxide?

9 Calculate the rate at which sulphate of potash (50 per cent K_2O) should be applied to supply:

(a) 50 kg/ha
(b) 85 kg/ha.

10 The tree stock production unit is to be moved. The soil has been sampled and the analysis results indicate the following requirements:

N: 100 kg/ha (N)
P: 25 kg/ha (P_2O_5)
K: 200 kg/ha (K_2O)
Mg: 75 kg/ha (Mg)

How much of the following fertilizers would be needed?

(a) Ammonium nitrate (33.5 per cent N)
(b) Superphosphate (20 per cent P_2O_5)
(c) Sulphate of potash (50 per cent K_2O)
(d) Kieserite (17 per cent Mg).

11 The annual top dressing requirement of the rose beds as reported in the *Daily Sail* gardening section is as follows:

N: 100 kg/ha (N)
P: 37 kg/ha (P_2O_5)
K: 65 kg/ha (K_2O)
Mg: 41 kg/ha (Mg).

The following fertilizers are available for use. How much of each is required?

(a) Ammonium nitrate (33.5 per cent N)
(b) Triple superphosphate (46 per cent P_2O_5)
(c) Sulphate of potash (50 per cent K_2O)
(d) Kieserite (17 per cent Mg).

12 50 kg/ha of magnesium is required for a seed bed preparation.

(a) Which magnesium fertilizer would you use?
(b) At which rate would you apply it?

13 After sampling and analysing a nursery soil, it is found that fertilizer is required at the following application rates:

N: 180 kg/ha
P_2O_5: 90 kg/ha
K_2O: 90 kg/ha.

Choose a suitable fertilizer from the list to use, and state its rate of application:

Fertilizer A: 8:16:16
Fertilizer B: 20:10:10
Fertilizer C: 24:18:18

Conclusions

1 Explain why potassium fertilizers are easy to apply as a foliar feed, whereas phosphorus ones are not.
2 State how magnesium deficiency may be overcome.
3 One 50 kg bag of fertilizer contains 15 per cent nitrogen, 15 per cent phosphorus (P_2O_5), and 20 per cent potassium (K_2O). How many bags of this fertilizer are required to supply 75 kg N, 75 kg P_2O_5, and 100 kg K_2O per hectare?
4 Name four fertilizers commonly used as base dressings.
5 State the role of nitrogen in the plant and the precautions to be taken when using nitrogenous fertilizers.

6 Which one of the following is a phosphorus fertilizer:

(a) Basic slag
(b) Nitrate of potash
(c) Hoof and horn
(d) Urea?

Exercise 20.7 *Area and amount of fertilizer required*

Background

Most horticulturalists need to know how much fertilizer to get out of the fertilizer store to complete the application recommended without any waste, either in terms of chemicals, or of time spent returning to the store to get additional quantities of fertilizer or returning surplus.

For this we need to know the 'area' of land that is to be treated with fertilizer. This is calculated simply by multiplying the length by the width of the area to be treated. For example:

(a) Tree production unit is 20 m by 50 m

$$\text{Area} = 20\,\text{m} \times 50\,\text{m} = 1000\,\text{m}^2$$

(b) Alstroemeria beds measure 2 m by 30 m

$$\text{Area} = 2\,\text{m} \times 30\,\text{m} = 60\,\text{m}^2$$

Aim

To increase ability in calculating areas for fertilizer application rates.

Method

Calculate the areas of the following units.

Results

1 AYR chrysanthemum beds measuring 4 m by 30 m
2 Polytunnel measuring 15 m by 20 m
3 Efford sand beds measuring 8 m by 15 m
4 Pot chrysanthemum benches measuring 25 m by 40 m
5 Ericaceae stock beds measuring 1.5 m by 8 m.

Conclusion

Explain the relationship between the area of land management and fertilizer recommendations.

Exercise 20.8 *Total fertilizer quantity measurements*

Background

Using the area measurement (m^2) (see Exercise 20.7) and the fertilizer application rate (g/m^2) (see Exercise 20.6), we can now calculate the total quantity of fertilizer required using the following formula:

Total quantity (Q) of fertilizer required = Area (m^2) × Application rate (g/m^2)

Example

A base dressing fertilizer has a rate of application of 75 kg/ha. How much fertilizer would be needed for a seed bed measuring 20 m by 50 m?

Step A   Area   = 20 m × 50 m = 1000 m^2

Step B   Quantity Q = 1000 m^2 × 7.5 g/m^2

              = 7500 g

              = 7.5 kg

Method

Calculate the total quantity of fertilizer required in the following situations.

Results

1 A base dressing of 75 kg/ha N for seed bed production of celery in a polytunnel measuring 20 m by 50 m. How much fertilizer is needed?
2 A base dressing of 150 kg/ha is recommended for a glasshouse measuring 40 m by 50 m. How much fertilizer is required?
3 Phosphorus pentoxide is required at 75 kg/ha. How much is needed for an *Alstroemeria* bed measuring 2 m by 15 m, using:

   (a) single superphosphate (18 per cent P_2O_5)?
   (b) triple superphosphate (46 per cent P_2O_5)?

4 A garden border measures 6 m by 20 m. Soil analysis recommendations state a required base dressing of :

   (a) Ammonium nitrate at 70 g/m^2
   (b) Triple superphosphate at 150 g/m^2
   (c) Sulphate of potash at 350 g/m^2
   (d) Kieserite at 240 g/m^2

How much of each of these fertilizers is needed?

Conclusions

(Covering all exercises in this chapter.)

1 State the role of nitrogen, phosphorus, potassium and magnesium in plant growth. Describe the symptoms of their deficiency in a named plant.

2 Explain the purpose of using fertilizers in horticulture.
3 Distinguish between the following terms;

(a) base dressing
(b) top dressing
(c) foliar feed
(d) liquid feed.

4 State the chemical units that fertilizer recommendations are quoted in for:

(a) nitrogen
(b) phosphorus
(c) potassium
(d) magnesium.

5 Describe how the method of nutrient release varies between fast and slow.
6 Explain the principle of nutrient release from controlled release fertilizers.
7 Name two straight fertilizers that supply phosphorus and potassium.
8 Describe two advantages and two disadvantages of using organic fertilizers compared to inorganic fertilizers.
9 In which one of the following soils is potassium deficiency most likely to occur?

(a) badly drained
(b) heavy textured
(c) sandy textured
(d) well drained.

Part Three   Pest and disease

21 Fungi

Background

The non-green plants such as bacteria and fungi, do not normally contain chlorophyll and have to obtain their nourishment from other organisms. This type of plant is known as a **parasite**. They are a major cause of disease, infecting plants in all areas of horticulture. However, a knowledge of their biology can aid horticulturalists in the timing of control measures.

Fungi are the 'decomposers' in the food chain. They help maintain soil fertility by recycling nutrients and decomposing organic matter. They secrete digestive enzymes onto the material which, once soluble, is absorbed into the fungi. They are cryptogamous organisms which, because they lack chlorophyll, feed on organic matter. (A cryptogam is a plant that reproduces by spores.)

There are over 80 000 different species of fungi; the largest is the puff ball and the smallest are single-celled organisms such as yeast. They may be saprophytic (feeding on dead and decaying matter) or parasitic (e.g. bracket fungi and Athlete's Food, Phytophthora, Pythium, Rhizoctonia, Fusarium and Botrytis).

Saprophytes live on dead and decaying material

Obligate saprophytes only use dead material

Facultative saprophytes prefer dead matter, but can use living material

Parasites derive their nutrition from living materials

Obligate parasites can only use live food (e.g. powdery and downy mildew, rusts, smuts). If you clear the site of live food you will kill the parasite. Powdery mildew, like other obligate parasites, weakens the plant but does not kill it.

Facultative parasites prefer live food, but can survive on dead. For prevention horticulturalists must destroy all dead plant material as well as living waste. For example honey fungus can live on dead matter for over twenty years before moving onto live food. Hygiene is therefore essential – always remove as much dead material as possible.

Common fungi include the mould fungi which grow on stale bread and cheese. Other examples of fungi include mucor, rust fungus, mushrooms, toadstools, puff balls and bracket fungi.
Fungi can be classified into four groups (see Table 21.1).

Class	Example fungi and diseases
Primitive fungi (Phycomycetes) (algae-like fungi)	Club root (*Plasmodiophora brassicae*) Damping off (*Pythium, Phytophthora*) Downy mildew (*Plasmopara viticola , Perenospora, Bremia*) Potato blight (*Phytophthora infestans and parasitica*)
Intermediate fungi (Ascomycetes) (sacrophytic fungi, over 30 000 species)	Powdery mildew (*Erisyphe*) Leaf spots (*Cercospora, Pseudopeziza*) Stem rots (*Didymella, Nectria*) Wilts (*Fusarium, Verticillium*) Dutch elm disease (*Ceratocytis ulmi*) Brown rot (*Monilinia fructigena*) Black spot (*Diplocarpon rosae*) Apple scab (*Venturia inaequalis*) Ergot (*Claviceps purpurea*)
Imperfect fungi (Deuteromycetes) (over 20 000 species). These fungi lack the ability for sexual reproduction, but are very similar in other respects to ascomycetes	*Penicillium chrysogenum* – an antibiotic (penicillin) *Dactylaria* – traps and feeds on passing nematode worms *Aspergillus* – a mould *Tichophyton* – athlete's foot *Rhizoctonia* – root rots *Fusarium spp* – wilts
Advanced fungi (Basidiomycetes) (over 25 000 species)	Rusts (numerous, e.g. *Puccinia graminis*) Smuts (numerous, e.g. *Ustilago avenae*) Bracket fungi Mushrooms and toadstools Puff balls

Table 21.1
Fungus classification

Beneficial effects of fungi

Saprophytes – These recycle nutrients – they are able to break down complex organic material and release valuable elements including nitrogen, carbon and sulphur. Fungi are very important in the carbon cycle. Approximately 50 per cent by weight of leaves is carbon. This is released into the atmosphere during decomposition as CO_2 and is therefore available to plants for photosynthesis.

Biotechnology – Brewing and baking, steroids, cheese making.

Antibiotics – The production of chemicals that inhibit the growth of competitors, e.g. Penicillium.

Weed control – Some fungi attack weeds including wild oats and cleavers. These are called mycoherbicides, e.g. *Colleotrichum gloeosporiodes aeschynomene*.

Pest control – e.g. *Verticillium lecanii* used to control glasshouse pests.

Control

A fungicide is a chemical control to kill fungi. Soil fungicides are normally applied on nurseries for prophylactic or preventative control, inhibiting the spread of mycelium in the media.

The thermal death point (TDP) is the temperature required to kill an organism. Most fungi have a TDP of 50°C for ten minutes and a few of 80°C for ten minutes. These parameters are considered with reference to soil sterilization procedures.

Exercise 21.1 *Mycelium investigation*

Background

Fungi are not made up of cells but of thread-like hyphae (singular hypha) which contain several nuclei, branching into a network structure. Sometimes cross-walls divide the hyphae. The hyphae spread throughout the soil penetrating between the particles. This network is referred to as mycelium (fungal mycelium). Mycelium are what gives a woodland soil its so called 'earthy' smell. It is believed that fungal mycelium constitutes the largest single living organism in the world and in the redwood forests of North America some have been estimated to extend several kilometres in diameter.

In this exercise fungi specimens (mycelium and spores) have been grown on agar gel, although stale bread, fruit and moist food are equally suitable. Small samples can then be transferred to a microscope slide and examined under a microscope.

Aim

To become familiar with the structure of fungal mycelium.

Apparatus

monocular microscopes	slides
needles	cover slips
white tiles	scalpels
Bunsen burner	petri dish cultures
cotton blue droppers	

Method

Select a petri dish containing fungal cultures of Botrytis, Alternaria and Penicillium (or alternative cultured fungi). For each sample complete the following:

1 Take a microscope slide.
2 Use a needle to lift of some of the fungal strands and place on the slide.
3 Place some cotton blue lacto-phenol dye over the slide (sometimes the dye is not necessary).
4 Using two needles carefully separate the strands to make them more clearly visible.
5 Lower a small glass cover slip slowly on top of the droplet using a needle for support.
6 Check that the microscope is on low power.
7 Move the slide around until the fungus is in vision and focus.
8 Move to high power by twisting the lens, looking alongside the microscope to ensure that the lens doesn't break the slide.
9 Focus on the fungus.
10 Draw a labelled diagram of your observations including:

   (a) colour of hyphae
   (b) shape of hyphae
   (c) presence/absence of cross walls
   (d) How spores are held in the hyphae
   (e) shape of spores
   (f) colour of spores.

11 Sterilize your needles in a Bunsen flame between each fungus.

Results

For each fungus draw diagrams of the following:

1 mycelium strands
2 hyphae showing:

   (a) colour of hyphae
   (b) presence/absence of cross walls

3 Spores showing:

   (a) how spores are held in the hyphae
   (b) colour of spores

Conclusions

1 When something goes mouldy what is actually happening to it?
2 Explain the terms:

(a) hyphae
(b) saprophyte
(c) parasite
(d) mycelium.

3 State five diseases causes by fungus.
4 Draw a labelled diagram to show a small portion of a fungal hyphae.
5 In what ways do fungi affect the natural world?

Exercise 21.2 *Spore cases and germinating spores*

Background

Fungi reproduce not by seed but from spores. The spores are supported on a special extended stalk tip of vertical hyphae called a sporangia. The spores are encased in a sporangium and are released into air currents when its wall breaks down. The ascomycete and basidiomycetes fungi groups reproduce sexually and their spore case is called a perithecia, but otherwise they fulfil the same function as a sporangium. Once the spores have been transported to a new site such as a leaf blade, they germinate and grow into the host material forming new mycelium fibres.

Fungi are very fast growers. For example it has been estimated that in twenty-four hours a half-tonne cow will manufacture 0.5 kg of protein, but half a tonne of yeast will grow to 50 tonnes and only use up a few square metres of soil. For this exercise spores may be collected from any suitable fungus and prepared on a microscope slide.

Aim

To investigate the method of fungus reproduction.

Apparatus

microscopes
slides of germinating spores

Method

1 Place the slide under the microscope and focus.
2 Squash the spore case gently with a needle to encourage spore emergence.
3 Draw a labelled diagram of your observations.

Results

Draw a diagram of the emerging/germinating spores.

Conclusions

1 How large are the spores?
2 How many spores are produced?
3 Name two types of spore casing produced by fungi.

Exercise 21.3 *Spore cases and spore survival techniques*

Background

Some spores are able to survive over winter until conditions for germination are improved. This normally involved having a harder resistant casing (sclerotia) as in botrytis. Using this knowledge horticulturalists are able to apply preventative fungicides prior to spore germination.

Aim

To investigate spore survival techniques.

Apparatus

Botrytis sclerotia
microscope slides
microscopes

Method

1 Observe the sclerotia (spore case) of botrytis.
2 Draw a labelled diagram of your observations.
3 Try to break open the case.

Results

Draw a diagram of your observations.

Conclusions

1 Suggest two advantages the fungus would derive from producing sclerotia.
2 Explain why fungi are 'cryptogamous' organisms.
3 What role do fungus play in the decomposition of materials?
4 Explain the difference between a 'facultative' and 'obligate' parasite.

Exercise 21.4 *Rust disease*

Background

Several diseases can result from fungi including blight, mildew and rust. There have been several fungal epidemics including the Irish potato famine of the 1840s which was caused by the potato blight fungus (*Phytophthora parasitica*) favoured by high temperatures and humidity. Dutch elm disease is also caused by a fungus carried by beetles from tree to tree.

Rust is a common and easily identified fungal disease. Susceptible plants include, particularly, groundsel, plums, beans, carnations, pelargoniums and chrysanthemums. The mycelium causes the leaf to have a ragged appearance around a raised dark brown leaf spot containing spores that can be brushed off with your fingers.

Aim

Recognition of rusts.

Apparatus

rust-infected plant
microscope

Method

1 Observe the specimen under the microscope.
2 Draw a labelled diagram of your observations.

Results

Draw a labelled diagram of your observations.

Conclusions

1 Which group of fungi does rust belong to?
2 List some plants susceptible to rust disease.
3 Which part of the plant does rust fungus normally attack?
4 Name four fungal diseases from the ascomycetes group.

Exercise 21.5 *Powdery mildew disease*

Background

These intermediate fungi infect leaves. Susceptible plants include particularly cucumber, roses and apples, although many country hedgerows also appear to be good hosts. The fungus sits on the upper surface of the leaf and inserts short feeding probes through which food is sucked up. They favour hot dry conditions and are particularly important because unlike most other fungi they do not require a wet surface to begin infecting.

Aim

Recognition of mildew.

Apparatus

mildew infected plant
microscope

Method

1 Observe the specimen under the microscope.
2 Draw a labelled diagram of your observations.

Results

Draw a labelled diagram of your observations.

Conclusions

1 Which group of fungi does mildew belong to?
2 List some susceptible plants.
3 What conditions stimulate powdery mildew growth?
4 How is powdery mildew different from most other fungi?
5 Name four fungal diseases from the primitive fungi group.

Exercise 21.6 *Structure of a mushroom*

Background

The advanced fungi reproduce by spores that are broadcast from a specialized hyphae called a 'fruiting body' such as a mushroom. Many people think that a mushroom is an individual plant, but in reality it is only part of a much more extensive mycelium organism living in the soil. Bracket fungi, such as Chicken-in-the-wood, which grow outwards from trees reveal to the trained horticulturalist that the whole tree will be infected by mycelium and will be slowly dying. When ripe, the gills of the mushroom turn brown and the spores are carried away in the wind.

Aim

To investigate the fruiting body structure.

Apparatus

microscope
parasol mushroom

Method

1 Observe the fruiting body.
2 Draw a labelled diagram.
3 Remove the stalk.
4 Observe the gills under the microscope and report any significant observations.

Results

Draw a diagram of the fruiting body and note your observations.

Conclusions

1 Suggest why toadstools may be found growing in dark areas of woodland where green plants cannot survive.
2 Explain what is meant by the term 'fruiting body'.
3 Which group of fungi are responsible for the production of 'fruiting bodies'.
4 State two environmental conditions that would favour the establishment of a fungus spore on a leaf surface.
5 Name two fungal diseases, the plants they attack and the symptoms they produce.

Exercise 21.7 *Spore mapping*

Background

The field mushroom (*Agaricus campestris*) is estimated to produce 100 000 spores per hour. Each spore is only 0.001 mm in diameter.

The mushroom itself is only the fruiting body of a much larger organism forming the mycelium living underground. There are over 80 000 species of fungus and identification (mycology) is a major science in itself. Often many species can only be separated by reference to a spore map. This is produced by placing a fungus fruiting body, gills downward, on a piece of white card and observing the resulting spore pattern and colouring that develops as the gills pump out spores.

Aim

To produce a spore map of a fungus.

Apparatus

fungus fruiting body
white paper

Method

1 Label the top right-hand corner of the paper with specimen details.
2 Remove the stalk from a mushroom with dark brown gills (i.e. mature).
3 Place the cap on a piece of paper, gills facing downwards.
4 After forty-eight hours discard the fungus; a spore map will have been produced.
5 Collect your map for reference.

Results

Paste your spore map into a notebook. This may be preserved for future reference by covering with transparent sticky plastic.

Conclusions

1 What can you conclude about the rate of production of spores?
2 In what way would your map have differed if left for longer?
3 What advantage does a fungus yield from producing an abundance of spores?
4 State four positive benefits of fungi to horticulturalists.

22 Insects and mites

Background

Insects and mites can be pests of horticultural crops, normally by feeding on the plant as a source of nutrition. The animal kingdom is enormous and it would be impossible to address every single pest. However, these exercises are designed to provide an introduction to pest recognition and management through investigating the life cycles of different organisms, in addition to understanding the damage done by such pests to plants, together with some chemical controls.

Typical pests of horticultural plants, which come mostly from the order anthropoda, include some crustaceans (e.g. woodlice), arachnida (e.g. scorpions, mites, daddy-long-legs), diplopoda (millipedes), chilpoda (centipedes) and insecta (insects).

Of these the insects are the most important. They may be divided simply into winged (pterygota) or wingless (apterygota) insects. Wingless pests include organisms such as springtails and bristletails. The winged insects are easily the largest group of pests and include organisms such as most bugs, beetles, butterflies, moths, ants, wasps, lace-wings and thrips.

Generally the damage that these organisms do to plants can be grouped into whether it is by gnawing, biting or sucking. For example, the thrip (*Thysanoptera*), otherwise known as a thunder bug or gnat, has a long feeding probe which it inserts into the petals of flowers and feeds on the sap. As it does so, it causes air to rush into the cell creating a characteristic silver-white streaking on the leaves and petals. For further symptoms of pest damage see Figure 24.11.

Specimen collection

Often, it is a particular phase in the life cycle of a pest which causes damage to plants. It will be useful to collect all life cycle

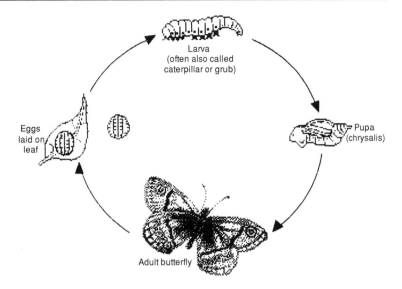

Figure 22.1
Typical pest life cycle stages

stages, including grubs and adults (see Figure 22.1). For example, it is the larvae stage (leatherjackets) of daddy-long-legs (cranefly) that feeds close to the surface of the soil damaging roots and underground parts of the stem of grasses.

In certain cases it may be useful to collect several pupae (Chrysalides, singular chrysalis) and allow one to hatch, so that a complete life cycle specimen collection can be built up.

Identification will also be aided by reference to a simple key. Organisms are often characterized by their number of legs and other easily observed features.

In this chapter the following pests will be investigated: vine weevils, wireworms and click beetles, leatherjackets and cranefly, cabbage root fly, cutworms, common gooseberry sawfly, cabbage white butterfly, red spider mite and aphids. In addition chemical insecticide controls will also be stated through reference to *The UK Pesticide Guide* (the 'green book'; Whitehead, 1995), previously introduced in Chapter 4.

Exercise 22.1 *Pit fall trapping*

Background

Despite their abundance it can be a difficult task to collect pests for investigation. Two methods are presented to enable this to be done. First, the creation of a pit fall trap. Second, bug hunting with a collecting jar called a pouter (see Exercise 22.2). In addition, rummaging through compost and old heaps often yields plenty of 'beasties' for study.

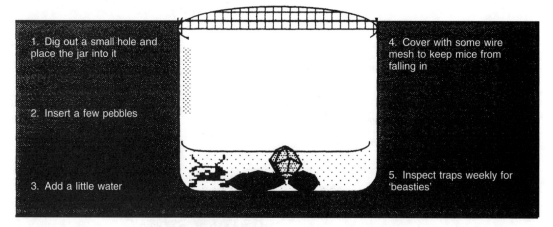

1. Dig out a small hole and place the jar into it

2. Insert a few pebbles

3. Add a little water

4. Cover with some wire mesh to keep mice from falling in

5. Inspect traps weekly for 'beasties'

Figure 22.2
Typical pit fall trap

The use of a pit fall trap is especially convenient since it requires only a little work to make and install. It involves inserting a plastic container into the ground into which beasties, in their inquisitiveness, fall. A clean half-pint plastic beer glass has been found to be ideal for this purpose. The insects are drowned in a little water in the base of the container and collected weekly for identification. This can be continued on a permanent basis throughout the year so that different life cycles are collected and also helps, to some degree, in assessing pest numbers.

Aim

To enable common pests to be pit fall trapped for study and private collections.

Apparatus

plastic container
a few pebbles
water
wire mess

Method

Set up the apparatus as shown in Figure 22.2.

Results

Record the type, and number of organisms found by reference to a key.

Conclusions

1 What sort of organisms are pests in horticulture?
2 Describe typical insect feeding methods that may damage plants.
3 Explain the difference between a 'larva' and a 'pupa'.

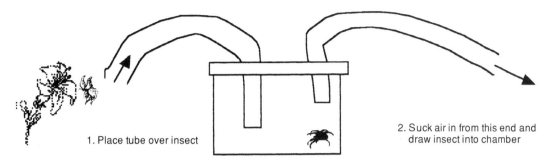

1. Place tube over insect

2. Suck air in from this end and draw insect into chamber

Figure 22.3
Typical home-made pouter

Exercise 22.2 *Pouter bug hunting*

Background

A pouter is a device used to collect organisms without needing to touch them directly (see Figure 22.3). Many insects are difficult to catch and a pouter is often useful to trap them before they have chance to escape. Old compost and the leaves of most trees and shrubs are a good source of insects.

Aim

To investigate and collect pests found feeding using a hand-held pouter.

Apparatus

pouter

Method

1 Rummage through old compost or plant material and collect any bugs found using a pouter.
2 Identify the insects using a suitable key.

Results

Identify and record the type (by reference to a key), and number of organisms found.

Conclusions

1 What sort of insects were mostly found feeding on old compost?
2 What sort of insects were mostly found feeding on leaves?
3 Explain what is meant by the term 'life cycle'.
4 State the life cycle stages of the 'beasties' that you trapped.

Exercise 22.3 *Vine weevils* (Otiorhynchus)

Background

These large to medium-sized flightless weevils are found throughout Britain and often called wingless weevils. Generally they damage the roots (grubs), leaves and stems (adults). Plants favoured for attack include pot plants under glass, ornamental plants in containers, cyclamen, strawberries, hops, azalea, rhododendron, camellia and almost any other outdoor grown plant.

Aim

Investigation of common horticultural pests.

Apparatus

> larvae, pupae and adult vine weevil
> *The UK Pesticide Guide* (Whitehead, 1995)

Method

1 Observe the specimens.
2 Draw the life cycle stages.

Results

1 State what damage to plants each of the life cycle stages causes and whether by biting or sucking.

    (a) larvae
    (b) pupae
    (c) adult

2 How does the snout differ from other beetles?
3 How long are the adult beetles?
4 When do the females lay their eggs?
5 In which month do the larvae start to pupate?

Conclusions

Using *The UK Pesticide Guide*, state what chemicals can be used to control the pest. Give the following details:

(a) active ingredients
(b) product names
(c) manufacturing company.

Exercise 22.4 *Wireworms and click beetles* (Agriotes, Athous and Ctenicera)

Background

Wireworms are probably the best-known soil inhabiting plant pests. Click beetles are named from the sound (click) they make as they flick themselves into the air to right themselves after falling on their back. About sixty species are found in Britain, the larvae of which are called wireworms.

Susceptible plants include grass, cereals, potatoes, sugar beet, brassicas, dwarf French beans, lettuce, onions, strawberries, tomatoes and anemones.

Aim

Investigation of common horticultural pests.

Apparatus

> larvae
> adult
> *The UK Pesticide Guide* (Whitehead, 1995)

Method

1 Observe the specimens.
2 Draw the life cycle stages.

Results

1 State what damage to plants each of the life cycle stages causes, and whether by biting or sucking:

(a) larvae
(b) pupae
(c) adult

2 When are the adult beetles formed?
3 When are the eggs laid in the soil?
4 How many pairs of legs have wireworms got?

Conclusions

Using *The UK Pesticide Guide*, state what chemicals can be used to control the pest. Give the following details:

(a) active ingredients
(b) product names
(c) manufacturing company.

Exercise 22.5 *Leatherjackets and cranefly* (Diptera, Tipula)

Background

The sworn enemies of turf managers, cranefly larvae are known as leatherjackets and thrive especially in prolonged damp weather in late summer. Their presence is often evident from a mass of birds, especially rooks, feeding on the grass.

Susceptible plants include grass, clover, strawberries, cereals, sugar beet, brassicas, courgettes and herbaceous garden plants.

Aim

Investigation of common horticultural pests.

Apparatus

all life cycle stages
The UK Pesticide Guide (Whitehead, 1995)

Method

1 Observe the specimens.
2 Draw a labelled diagram of the cranefly structure.

Results

1 State what damage to plants each of the life cycle stages causes and whether by biting or sucking:

(a) larvae
(b) pupae
(c) adult

2 What are cranefly also known as?
3 How many legs do leatherjackets have?

Conclusions

Using *The UK Pesticide Guide*, state what chemicals can be used to control the pest. Give the following details:

(a) active ingredients
(b) product names
(c) manufacturing company.

Exercise 22.6 *Cabbage root fly* (Delia radicum)

Background

A destructive pest of Brassicaceae (previously Cruciferae family) plants the larvae attack seedlings and mature plants, damaging the root system.

Susceptible crops are cauliflowers, cabbage, Brussels sprouts, calabrese, Chinese cabbage, radish, swede, turnip, garden stocks and wall flowers.

Aim

Investigation of common horticultural pests.

Apparatus

life cycle stages: larvae, pupae and adult
The UK Pesticide Guide (Whitehead, 1995)

Method

1 Observe the specimens.
2 Draw a labelled diagram of the life cycle.

Results

1 State what damage to plants each of the life cycle stages causes and whether by biting or sucking:

(a) larvae
(b) pupae
(c) adult.

2 What is the colour of the adult fly?
3 How long is the fly?
4 When can the first eggs be found in plants?
5 What colour are the maggots?
6 How many legs have the maggots?

Conclusions

Using *The UK Pesticide Guide*, state what chemicals can be used to control the pest. Give the following details:

(a) active ingredients
(b) product names
(c) manufacturing company.

Exercise 22.7 *Cutworms*

Background

Cutworm is the name given to caterpillars of moths which feed on plants at ground level often severing them from the tap root. Moths which produce caterpillars known as 'cutworms' include the Turnip moth (*Agrotis segetum*), Garden dart moth (*Euxoa nigricans*), Whiteline dart moth (*Euxoa tritici*) and the Yellow underwing moth (*Noctua pronuba*).

Susceptible plants include lettuce, beet, celery, carrots, leeks, potatoes, turnips, swedes and other plants with tender tap roots.

Aim

Investigation of common horticultural pests.

Apparatus

life cycle stages
The UK Pesticide Guide (Whitehead, 1995)

Method

1 Observe the specimens.
2 Draw a labelled diagram of the life cycle.

Results

1 State what damage to plants each of the life cycle stages causes and whether by biting or sucking:

(a) larvae
(b) pupae
(c) adult.

2 How long are the larvae?
3 What colour are the larvae?
4 How many legs do the larvae have?

Conclusions

Using *The UK Pesticide Guide,* state what chemicals can be used to control the pest. Give the following detail:

(a) active ingredients
(b) product names
(c) manufacturing company.

Exercise 22.8 *Common gooseberry sawfly* (Nematus ribesii)

Background

The caterpillars of this sawfly feed on the leaves of red and white currants and gooseberries. The rose-leaf rolling sawfly (*Blennocampa pusilla*) attacks leaves of rose plants.

Aim

Investigation of common horticultural pests.

Apparatus

life cycle stages
The UK Pesticide Guide (Whitehead, 1995)

Method

1 Observe the specimens.
2 Draw a labelled diagram of the life cycle.

Results

1 State what damage to plants each of the life cycle stages causes and whether by biting or sucking:

(a) larvae
(b) pupae
(c) adult.

2 How many pairs of wings do the adults have?
3 What colour are the larvae?
4 How long are the larvae?

Conclusions

Using *The UK Pesticide Guide*, state what chemicals can be used to control the pest. Give the following details:

(a) active ingredients
(b) product names
(c) manufacturing company.

Exercise 22.9 *Cabbage white butterfly* (Mamestra brassicae)

Background

The butterfly caterpillars feed on the aerial parts of most brassicas, almost all Brassicaceae (previously Cruciferae) plants, garden nasturtiums and general allotment plants, causing severe defoliation.

Aim

Investigation of common horticultural pests.

Apparatus

life cycle specimens
The UK Pesticide Guide (Whitehead, 1995)

Method

1 Observe the specimens.
2 Draw a labelled diagram of the life cycle.

Results

1 State what damage to plants each of the life cycle stages causes and whether by biting or sucking:

(a) larvae
(b) pupae
(c) adult.

2 When do the females lay their eggs?
3 How long are the larvae?
4 What colour are the larvae?
5 In which month do the larvae start to pupate?

Conclusions

Using *The UK Pesticide Guide*, state what chemicals can be used to control the pest. Give the following details:

(a) active ingredients
(b) product names
(c) manufacturing company.

Exercise 22.10 *Red spider mite* (Tetranychus urticae)

Background

Also known as the two-spotted spider mite and rarely looking red, this is a pest on protected crops but can damage outdoor plants in hot dry summers. The mites feed on the underside of leaves sucking the sap and also spinning a fine silken web over the plant. The females turn red after mating in the autumn. They may be controlled biologically using another mite called *Phytoseiulus persimilis*, which is in fact a lot redder than red spider mite itself (see Chapter 24).

Susceptible plants include all glasshouse crops, strawberries, hops, beans, black currants, cane fruit, hops, fruit trees and ornamentals.

Aim

Investigation of common horticultural pests.

Apparatus

infected plants
binocular microscope
The UK Pesticide Guide (Whitehead, 1995)

Method

1 Observe the specimens under the microscope.
2 Draw a simple diagram of the mite's life cycle.

Results

1 How many legs have the mites?
2 How big are they?
3 State the damage to plants caused by the adult life cycle stage and whether by biting or sucking:

(a) larvae
(b) pupae
(c) adult.

Conclusions

1 Using *The UK Pesticide Guide*, state what chemicals can be used to control the pest. Give the following details:

(a) active ingredients
(b) product names
(c) manufacturing company.

2 State the predator often used in biological control of this pest.
3 Explain what is thought to cause the females to turn red.

Exercise 22.11 *Aphids* (Hemiptera)

Background

Aphids occur in a tremendous number of varieties although we normally consider them as just one group. The green-coloured aphid only appears in midsummer. Injury causes leaves to curl, young shoots twist and fruit is small and premature.

Aim

Investigation of common horticultural pests.

Apparatus

> aphid-infected plants
> binocular microscope
> *The UK Pesticide Guide* (Whitehead, 1995)

Method

1 Observe the specimens under the microscope.
2 Draw a simple diagram of the insect's life cycle.

Results

1 How many legs have the aphids?
2 How big are they?
3 State what damage to plants each of the life cycle stages causes and whether by biting or sucking:

(a) larvae
(b) pupae
(c) adult.

Conclusions

Using *The UK Pesticide Guide*, state what chemicals can be used to control the pest. Give the following details:

(a) active ingredients
(b) product names
(c) manufacturing company.

23 Nematodes

Background

Nematodes, also called eelworms and roundworms due to their appearance, are plant pests that live in the soil and are normally encouraged by wet conditions. They help in the breakdown and recycling of organic matter in the soil to form humus.

Nematodes are unsegmented animals not closely related to ordinary worms. They have long thin thread-like bodies ('nema' is Greek for 'thread'). The largest worms are thought to live in Australia and can grow up to 3 m long. The *Ascaris* nematode lives in the intestines of pigs and grows up to 25 cm long. It is the largest nematode found in Britain, and may be deadly to man. Many are small or microscopic and occur in extremely large numbers especially in wet conditions. There are over 10 000 species of nematodes worldwide.

There are many types of nematodes that affect plants in the UK including:

beet cyst nematode (*Heterodera schachtii*)
pea cyst nematodes (*Heterodera goettingiana*)
cereal cyst nematode (*Heterodera avenae*)
chrysanthemum nematode (*Aphelenchoides ritzemabosi*)
leaf nematodes (*Aphelenchoides fragariae* and *Aphelenchoides ritzemabosi*)
stubby-root nematode (*Trichodorus* and *Paratrichodous* spp)
needle nematodes (*Longidorous* spp)
stem nematodes (*Ditylenchus dipsaci*)
root-lesion nematodes (*Pratylenchus penetrans*)
dagger nematodes (*Xiphinema diversicaudatum*)
potato cyst nematode (*Globodera rostochiensis* and *Globodera pallida*)
potato tuber nematode (*Ditylenchus destructor*)
root knot nematodes (*Meloidogyne* spp.).

These affect a vast range of plants from protected crops including tulips, strawberries, vegetables and some weeds. The main causes of infestation are growing a plant too frequently in the rotation and wet conditions. Several species of nematodes are parasites of garden pests such as slugs, and several commercial products are now available for purchase as biological control agents (see Chapter 24).

Common characteristics

Effects on crops

Low populations only damage a few roots therefore the horticulturalist may be unaware that the land is infected. As nematode populations increase so plant yields decrease.

Symptoms of attack

- small patch or patches in the crop
- stunted growth and unhealthy looking foliage
- outer leaves wilt in sunshine, turn yellow prematurely and die
- the heart leaves may remain green but are undersized
- roots remain stunted
- plants are easily removed from the soil
- tap root is small and short or absent
- development of 'hunger' roots (excessive lateral root development)
- small lemon-shaped nematodes can be seen protruding from dead roots

Method of spread

- soil movements, e.g. on tractors/machinery/animals
- carried on plants, e.g. seed potatoes/plants for transplanting
- strong winds blowing soil
- too frequent cropping of host plants

Control methods

- use resistant plant varieties
- biological control, e.g. parasitic fungi

- crop rotations may be the only economical method of control, e.g. crop only one year in three as a minimum, rising to one in five years on poor soils

Exercise 23.1 *Vinegar nematode* (Angiullula acetii)

Background

Nematodes are simple organisms containing a tube-like gut and separate mouth and anus. Typically they may be seen as small, white or transparent threads of cotton wool. They live by their millions in almost every habitat. It has been estimated that $1 m^2$ of woodland soil contains ten million nematodes. They move through the soil in films of water in a characteristic 'side-winder'-like motion caused by the contraction of four muscles which subsequently bend and twist the body.

The vinegar nematode causes a characteristic acid taste to vinegar and old beer. Upset tummies resulting from drinking outdated cloudy beer are sometimes caused through nematodes being passed into the stomach. They are an easy source of nematodes to collect and a useful starting point to aid identification.

Aim

To observe living male and female nematodes.

Apparatus

garden centre purchased nematodes kit or infected cider vinegar/ cloudy beer
microscope
0.2 per cent neutral red stain
filter paper
slides and cover slip
dropping pipette

Method

1 Add a drop of infected water solution to a microscope slide.
2 If desired, add one droplet of 0.2 per cent neutral red stain to the solution to aid clarity.
3 Observe the organisms under the microscope and note the speed with which they move.
4 Place a cover slip over the solution.
5 Place the filter paper against the edge of the cover slip to withdraw some water and slow down the movement of the nematodes.
6 Observe the different growth stages present.
7 State how many male and females are present.
8 What are the two main differences between the male and female nematodes?
9 How large is each nematode (estimate): 0.25 mm, 0.5 mm, 1 mm, 2 mm or 3 mm?

Results

Enter your observations in the table provided:.

Features of vinegar nematodes				
Estimate of sizes of nematodes present		Differences between males and females		Growth stages present
Number of males		1		1
		2		2
Number of females		3		3

Conclusions

1 Explain why some people experience an upset tummy after drinking real ale.
2 State a positive use of nematodes in horticulture.
3 Describe the symptoms of nematode attack on horticultural plants.

Exercise 23.2 *Nematode damage to plants*

Background

Nematode damage can remain undetected until severe plant losses occur. A list of the typical symptoms in given in the earlier background section and may be helpful to refer to here. A good source of materials are rotten bulbs and plants that has been thrown onto a compost heap.

Aim

To recognize symptoms of nematode attack in a range of plants.

Apparatus

> root knot nematode on cucumber and carnations
> chrysanthemum nematode on chrysanthemums
> narcissus stem bulb nematode
> stem and bulb nematode on onion

Method

Collect specimens of the above-listed plants and describe the damage caused in each.

Results

Enter your observations in the table provided.

Nematode	Symptoms of damage	
Root knot nematode (*Meloidogne* spp)	1	
1 Cucumber 2 Carnation	2	
Chrysanthemum nematode (*Aphelenchoides ritzemabosi*)	1	
1 Chrysanthemum	2	
Stem and bulb nematode (*Ditylenchus dipsaci*)	1	
1 Narcissus 2 Onion	2	

Conclusions

1 State how each of the above crop pests may be controlled or prevented.
2 List typical methods by which nematodes are spread.
3 Explain what is meant by the term 'parasitic nematode'.

Exercise 23.3 *Potato cyst nematodes* (Globodera spp.) *in soils*

Background

Potato cyst nematodes can be found in most soils but particularly those which have been used to grow potatoes. They result in damaged and stunted plants with patchy growth and low yields. The nematodes burrow into the plant and after fertilization the female swells up into a hard ball containing between 200 and 600 eggs. The egg case is called a 'cyst'. They can survive in the soil for up to ten years and are easily identified by their red-brown lemon-shaped pods. Sometimes they can be confused with weed seeds. Once the population size of this nematode has been determined an appropriate management practice can be selected and implemented to control the infestation.

In this exercise a sample of soil is mixed with water. The soil mineral fraction sinks as it is heavier than the cysts which will remain floating, along with other floating material, and is poured out (decanted) onto filter paper so that the cysts can be counted.

Aim

To ascertain the population size of potato cyst nematode cysts in a soil sample.

Apparatus

50 g air dry soil	white tray
1 litre conical flask	hand lens
large funnel	balance
tripod	stirring rod
beaker	weigh boat
filter paper	

Method

1 Weigh out approximately 50 g of air dried soil, ground to pass through a 2 mm sieve.
2 Transfer the soil to a 1 litre conical flask.
3 Make up to the 400 ml mark with tap water.
4 Swirl to thoroughly wet-up (hydrate) the soil.
5 Fill flask to the top using a trickle of water so that no floating material is spilt.
6 Fold the filter paper into four and place in the funnel in the tripod over the beaker (see Figure 23.1).

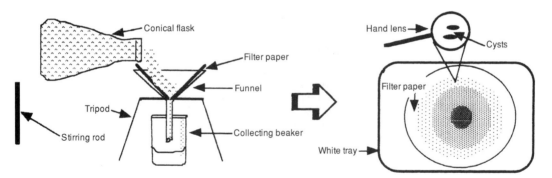

Figure 23.1
Apparatus for potato cyst nematode extraction

7 Pour the floating material into the funnel, simultaneously rotating the flask.
8 After decanting 5 cm of solution, refill the flask and continue as in step 7 until all the floating material has been decanted into the funnel.
9 Use the stirring rod to swirl the solution retained by the filter paper and encourage the floating material to move onto the filter paper.
10 After five minutes pierce the bottom of the filter paper and allow to drain.
11 Empty waste soil remaining in flask into wastes bin.
12 Transfer the filter paper onto a white tray.
13 Use a hand lens to count the number of cysts present (red-brown lemon shaped pods) (see Figure 23.1).
14 Do not confuse weed seeds with cysts.

Results

State your results in the table provided and select your management practice.

Number of cysts found in soil sample per 50g	
Number of cysts per 50g	Management practice
0	Can plant potatoes of any variety
1–5	Will probably get good crop with any variety
5–10	May get losses in patches with susceptible varieties
11–20	Will probably be all right with earlies or plant a resistant variety or use a soil nematicide
21–40	Leave land for two–four years
41+	Leave land for six years
Resulting management decision for sample soil	

Conclusions

1 How may the pest damage be recognized by looking in the field at growing plants?
2 How may the pest damage be recognized by looking at the roots?
3 How does the pest survive in the soil?
4 How many eggs may a cyst contain?
5 What has the cyst stage developed from?
6 How long may eggs survive in the soil?
7 How may a cyst be recognized with a hand lens?

24 Biological control

Background

The principle of biological control is the use of beneficial organisms (natural enemies) as **predators** or **parasites** to control common pests as an alternative to using chemicals.

A predator is simply an organism that consumes another providing energy, such as a fox eating a chicken. They may kill many pests during their life cycle. For example the mite called *Phytoseiulus persimilis* hunts and kills glasshouse red spider mite. Probably the best-known biological control using predators is the practice of introducing lady birds to a garden to feed on the greenfly population. Another type of tropical ladybird called *Cryptolaemus montruzieri* (see Figure 24.1) has been used very successfully in the control of mealybugs.

A parasite is an organism that feeds off of a living host pest but, in biological control, also results in the eventual death of the pest. Mostly, only one pest is killed during the life cycle of the parasite. Examples include the the use of a parasitic wasp called *Metaphycus helvolus* (see Figure 24.2) which lays its eggs in the juvenile stage of scale insects. As the wasp egg hatches the grub eats away at the protein-rich scale killing the pest.

Similarly, consider the increasing use of nematodes (see Figure 24.3) to control garden slugs and vine weevils. The nematode burrows into the slug's skin and eats it from within. A **pathogen** by comparison is normally a form of disease; examples include the use of bacteria and fungi to kill pest insects.

Several pests are 'notifiable' and if found should be reported to the Ministry of Agriculture, Fisheries and Food. Examples include Colorado beetle, plum pox, South American leaf miner, western flower thrip, and wart disease on potatoes.

Many natural enemies have been introduced to Britain from overseas. *Phytoseiulus*, for example, was first discovered in Germany in the 1960s on a consignment of orchids arriving from Chile.

Figure 24.1
Cryptolaemus montruzieri *adult feeding on mealybugs. Courtesy Koppert Biological Systems*

Figure 24.2
Metaphycus helvolus *parasitizing scale insect. Courtesy Dr Mike Copland, Wye College*

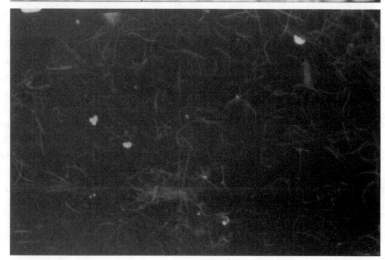

Figure 24.3
Heterorhabditis megidis *nematodes. Courtesy Koppert Biological Systems*

The use of biological control agents (natural enemies) is increasing. Many pests have grown immune/resistant to chemicals and it is a continuing contest to create effective pesticides. Every time a pesticide is used the plant's quality (colour, texture, consistency, etc.) has been shown to decrease. Approximately 1–2 per cent of yield may be lost every time chemicals are used. Biological control is now the standard pest control on salads grown under glass. There is also growing pressure against the use of chemicals in other food crops. For example, advice in dry summers has been to peel carrots as some of the outer skin may contain harmful pesticide residues. In addition, bore-hole water supply sampling has revealed that pesticide and herbicide residues are rising. Much research is being conducted to extend our knowledge in these areas and at reducing our dependence upon chemical control. Many glasshouse crops are grown under contract to supermarkets who stipulate that no chemical controls are to be used without their prior approval. 'Nature's Choice' produce is sold at a premium price because, as far as practicable, they have been grown without using chemicals. Pot plants and cut flowers, however, present more of a problem. Ideally, control by biological methods is desirable but the public are wary about buying plants with 'strange' insects (predators) crawling around on them. There is a great deal of consumer educating still to be done in this field.

Advantages of biological control

- Environmental benefits
- Increase in yield and quality
- No halt to plant growth (phytotoxicity)
- Permanent solution to pest problem
- Specific to pest
- Generally safe to other beneficial organisms
- Safer to operators and public
- No 'resistance' problems

Disadvantages of biological control

- Initially slow to establish and gain control
- Live organisms that sometimes die in transport
- Operators require knowledge of life cycle stages
- Pesticide and insecticides use restricted
- Correct environmental conditions are needed (e.g. light and humidity)
- May consume other natural enemies (e.g. *Macrolophus* eating *Phytoseiulus*)
- Less successful on outdoor plants due to dispersal and colder temperature
- Life forms are unpredictable and may be difficult to rear

In these exercises some aspects of biological control will be investigated as an introduction to an increasingly important area of horticulture. Example predators can be purchased from suppliers listed in trade magazines and see 'Suppliers' at the end of this book. Alternatively, most commercial horticulturalists and increasing numbers of garden centres stock some natural controls or can order biological control agents.

Exercise 24.1 *Common uses of biological control agents*

Background

Predators and parasites control pests in distinctly different ways. The following biological control agents include both types, but should be considered with reference to their practical use in horticulture. For example, the use of the parasitic wasp *Encarsia formosa* is only effective if introduced early at a density of one wasp scale per plant, when the whitefly pest problem is no greater than one whitefly per twenty plants. They also require a minimum temperature of 20°C, otherwise they do not fly around from plant to plant, and walking gives less effective control.

Tomato growers also benefit from a predator which can consume between ten and fifty whitefly per day. It is called *Macrolophus caliginosus* and has been used successfully in Europe. They have also been known to feed on spider mites, moth eggs and aphids. Used together with *Encarsia*, they can control whitefly more effectively. Obviously, no insecticides should be used as they are likely to kill the agent. Table 24.1 summarizes the breadth and type of natural enemy that can be used to control a range of insect pests.

Aim

To recognize beneficial insects and their method of controlling undesirable pests.

Apparatus

Various agents including:

> *Orius laevigatus, Bacillus thuringiensis, Verticillium lecanii* parasitized whiteflies, *Phytoseiulus persimilis* mite, nematode (*Phasmarhabditis hermaphrodita*) parasitized slugs and vine weevils, *Encarsia formosa* parasitized whitefly scales
> microscopes
> variety of hand lenses to test

Method

1 Observe the biological control agent under the microscopes.
2 Describe your observations (e.g. draw a diagram showing size and characteristics).
3 State which pest the natural enemy controls and whether predator, parasite or pathogen.
4 Repeat for other agents.

Table 24.1

Pest	Natural enemies	
Aphids	Aphelinus abdominalis	– parasitic wasp
	Aphidius colemani	– parasitic wasp
	Aphidoletes aphidimyza	– gall-midge
	Chysoperla carnea	– lacewing
	Delphastus pusillus	– predatory beetle
	Hippodamia convergens	– ladybird
	Verticillium lecanii	– pathogenic fungus
	Metarhizium anisopliae	– pathogenic fungus
Caterpillars	Bacillus thuringiensis	– pathogenic bacterium extract
	Trichogramma evanescens	– parasitic wasp
		Pheromone traps
Leaf-miners	Diglyphus isaea	– parasitic wasp
	Dacnusa sibirica	– parasitic wasp
Mealybugs	Cryptolaemus montruzieri	– predatory tropical ladybird
	Leptomastix dactylopii	– parasitic wasp
Red spider mite	Phytoseiulus persimilis	– predatory mite
	Therodiplosis persicae	– predatory mite
	Typhlodromus pyri	– predatory mite
	Amblyseius californicus	– predatory mite
Leaf-hopper	Anagrus atomus	– parasitic wasp
Scale insects	Metaphycus helvolus	– parasitic wasp
	Chilocorus	– predatory beetle
Sciarid flies	Steinernema carpocapsae	– parasitic nematode
	Hypoaspis miles	– predatory mite
Slugs	Phasmarhabditis hermaphrodita	– parasitic nematode
Thrips	Amblyseius mackenziei, A. barkeri	– predatory mite
	A. cucumeris, A. degenerans	– predatory mite
	Orius majusculus, O. spp	– predatory bug
	Verticillium lecanii	– pathogenic fungus
Vine weevils	Steinernema carpocapsae	– parasitic nematode
	Heterorhabditis megidis	– parasitic nematode
Whiteflies	Encarsia formosa	– parasitic wasp
	Macrolophus caliginosus	– predatory bug
	Delphastus pusillus	– predatory black beetle
	Verticillium lecanii	– pathogenic fungus

Results

1 Whitefly scales parasitized by the *Encarsia formosa* wasp (see Figure 24.4).

    (a) Describe it (e.g. shape, colour, legs, antennae, speed of movement/ flight).

    (b) Is this a parasite or a predator?

    (c) Which pests does it control?

    (d) List five other wasps and the pests that they control.

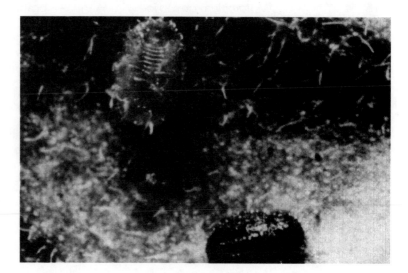

Figure 24.4
Whitefly scale unparasitized (white) and parasitized (black) by parasitic wasp Encarsia formosa. *Courtesy Koppert Biological Systems*

2 *Orius laevigatus*, an anthocorid bug (see Figure 24.5) (also investigate *O. insidiosus, O. tristicolor, O. albidipennis* and *O. majusculus*).

    (a) Describe it (e.g. shape, colour, legs, antennae, speed of movement/ flight).

    (b) Is it a parasite or a predator?

    (c) Which pests does it control?

    (d) What additional action should be taken if this particular pest is discovered on your plants?

    (e) List one other bug and the pest that it controls.

3 *Phytoseiulus persimilis*, mite (see Figure 24.6).

    (a) Describe it (e.g. shape, colour, legs, antennae, speed of movement/ flight).

    (b) Is it a parasite or a predator?

    (c) Which pests does it control?

    (d) List five other mites and the pests that they control.

Figure 24.5
Orius laevigatus *adult attacking adult thrip. Courtesy Koppert Biological Systems*

Figure 24.6
Phytoseiulus persimilis *attacking red spider mite. Courtesy Koppert Biological Systems*

4 Nematode (*Phasmarhabditis hermaphrodita*) parasitized slugs (see Figures 24.7 and 24.8) and vine weevils.

(a) Describe it (e.g. shape, colour, speed of movement).
(b) Is it a parasite or a predator?
(c) Which pests does it control?
(d) List two other species of nematode and the pests that they may control.

5 *Verticillium lecanii* fungus infected whiteflies (see Figure 24.9).

(a) Describe it (e.g. shape, colour, movement/flight, fungus on legs, hyphae visible, mouldy, fluffy).

Figure 24.7
Slug with swollen mantle after infection with nematodes. Courtesy Defenders Ltd

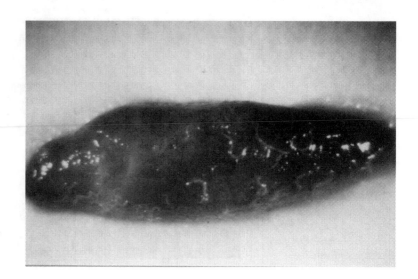

Figure 24.8
Slug killed and decomposing to release another generation of parasitic nematodes. Courtesy Defenders Ltd

(b) Is it a pathogen, parasite or predator?
(c) Which pests does it control?
(d) List two other pests that *Verticillium lecanii* may control.

6 *Bacillus thuringiensis* (a bacterium extract) infected caterpillars (see Figure 24.10).

(a) Describe it (e.g. shape, colour, movement, paralyzed, general appearance).
(b) Is it a pathogen, parasite, or predator?
(c) Which pests does it control?
(d) List two other biological controls for caterpillar pests.

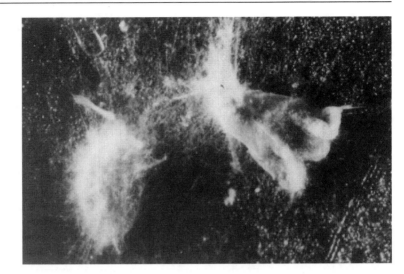

Figure 24.9
Whitefly adult and larvae infected by Verticillium lecanii. *Courtesy Koppert Biological Systems*

Figure 24.10
Caterpillar killed by Bacillus thuringiensis, *suspended from leaf by hind legs. Courtesy Koppert Biological Systems*

Conclusions

1 State the meaning of the term biological control.
2 Describe four precautions that need to be taken to ensure effective parasitism of whitefly by *Encarsia* wasp.
3 Explain the term 'parasite' as used in biological control.
4 Give an example of a parasite other than *Encarsia*, and the insect that it attacks.
5 Explain the term 'predator' as used in biological control.
6 Give an example of a predator and the pest that it preys on.
7 Give two examples of biological control of pests in horticulture.
8 Describe four other control organisms for use against named glasshouse or field pests.
9 Explain why the use of biological control agents is likely to increase.

Exercise 24.2 *Diagnosis of pest damage*

Background

Diagnosing plant disorders can be a challenging occupation. Unhealthy plants sometimes give rise to conditions that favour pest attack. Once a pest becomes established on a plant it may invite fungal diseases, feeding on honeydew and excrement, or viruses, introduced through the method of feeding. One thrip can infect fifty tomato plants with tomato spotted wilt virus (TSWV).

When the pest has been identified a suitable control may be selected. An essential aid to this section should be the purchase of a small pocket

Figure 24.11

Field guide of typical symptoms of plant damage by insect pests

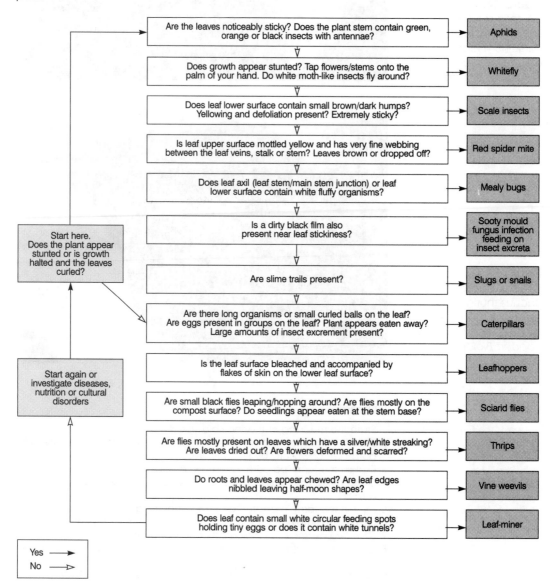

238

hand lens (×10 magnification) or similar magnifying glass. This can be used to identify both pests and symptoms of plant damage.

Problem pests can normally be separated into two groups. Those that damage the plant by biting/chewing and those that are sap-suckers. Earwigs, caterpillars, slugs and vine weevils, for example, all chew the foliage. The sap-suckers, including aphids and thrips, pierce the leaf and petals with their needle-like stylet and then suck out the nutrient-rich sap causing air to rush into the cells, collapsing them. In the case of thrips this leaves behind a characteristic silver/white streaking pattern. Leaf-miners may be recognized by the piercing dots left on the leaves. Similarly, snails may be recognized by the slime trail left behind. Figure 24.11 is a field guide of typical symptoms of plant damage by insect pests which may be a useful aid to describing the visual signs of pest damage.

Aim

To describe the visual symptoms of a range of pest damage.

Apparatus

hand lens
selection of problem plants
Figure 24.11

Method

1 Observe some problem plants attacked by a range of insect pests.
2 Describe the symptoms of the damage caused, using the field guide (Figure 24.11) as a pointer.

Results

Produce a table of results showing the types of symptoms found against the insect pest investigated.

Conclusions

1 Describe the different symptoms of pest attack between plants that damage by sucking and biting.
2 Why are unhealthy plants more likely to attract pest insects than healthy plants?
3 Describe why fungal diseases may become easily established on plants suffering from pest attack.

Exercise 24.3 *Monitoring pest levels*

Background

An essential aid to correct pest management is the use of monitoring devices to indicate correctly the population size of the problem pest. In short, detection is the basis of control. The sooner harmful organisms are detected, the more effective will be a biocontrol programme.

Locating the first pest insect can be difficult; 'trapping' can be a very important aid for this. For outdoor grown plants the use of a pitfall trap, as described in Chapter 22 is highly recommended. For glasshouse-grown plants the use of sticky traps is essential. These are advanced forms of fly paper to which the insects stick and can be subsequently identified and counted. There should be no such thing as an empty glasshouse; a number of pests are able to survive the period between crops and may be caught before they can damage new plants, thereby helping to make a clean start to the cropping programme. Glasshouses should always have sticky traps present by the doors and vents so that an accurate assessment can be made of pest levels through constant monitoring.

There are several traps available including white, orange, yellow and blue. These colours appear to be more attractive to insects, although other factors, such as glue type, size and shape have also been found to be important. White and orange traps are used for leaf-miner monitoring. The most extensively used traps are blue and yellow. The yellow traps are the most common and slightly cheaper, used to attract whitefly, leaf-miners, aphids and sciarid flies. The effectiveness of the traps are influenced by environmental factors including day length, light radiation and temperature which affect the behaviour pattern of the pest insects.

The blue traps are essential to detect thrip levels as thrips are not easily attracted to the yellow traps. The blue traps catch up to four times as many female thrips as the the yellow ones. Counts as low as four or five thrips per sticky trap should be considered dangerous levels. Chemical sprays may be needed to lower pest levels prior to the introduction of a natural enemy. Three simple principles should govern the use of these biological agents:

1 Introduce biological agents early in the season before pest levels rise.
2 Introduce biological agents at an appropriate density to control the pest (e.g. *Amblyseius cucumeris* at fifty predators per m^2 per week all year round to control thrips).
3 Adopt a whole-site approach. Good husbandry in one glasshouse may be counterproductive if pests can fly in through vents from another badly infected glasshouse.

There are instances where sticky traps may be used as a control. However, as a general rule, sticky traps are not a pest control method. They are a management tool useful to assess pest population levels accurately and devise suitable control programmes based upon the information that they yield.

Natural enemies are readily available to commercial growers and can be cheaper than chemicals because pests are increasingly immune. The use of biological agents themselves does not always run smoothly. Being live organisms they sometimes die in transport and do require the horticulturalist to have greater knowledge of the life cycles of both the agents and the pests than might otherwise be the case with chemical control. Generally horticulturalists should approach the subject through a balance of pest management methods to ensure effective control.

Aim

To use a range of techniques to detect and monitor pest insects.

Apparatus

   blue and yellow sticky traps
   any insect identification key/handbook

Method

1 Set up blue in one glasshouse and yellow sticky traps in a different glasshouse.
2 Observe the traps weekly and record the insect pests caught over the next three weeks.

Results

Enter your results in the table provided.

Observation period	Yellow sticky trap				Blue sticky trap			
Week 1 Insect pests detected								
Number of occurrences								
Week 2 Insect pests detected								
Number of occurrences								
Week 3 Insect pests detected								
Number of occurrences								

Conclusions

1 List five difficulties associated with biological control and explain why its use is likely to increase.
2 Describe the application of biological control organisms to a named crop with which you are familiar.
3 Name two organisms commonly used for biological control in glasshouse flower crops?
4 Explain the importance of pest monitoring and describe some methods to accumulate adequate information.
5 State the advantage of using blue traps rather than yellow traps for thrip monitoring.
6 Describe the environmental factors which influence the successfulness of using sticky traps.
7 For each of the insect species detected on the sticky traps, state a suitable beneficial organism which may be used as biological control agent.

Exercise 24.4 *Predators and prey*

Background

Pest management is a continuing battle of selecting the right weapon from the armoury to defeat the organism. Predators will often only lay eggs when pest numbers are causing plant losses. This is to ensure that there is sufficient prey for the hatched young to feed on. Ladybird numbers, for example, correspond closely with the population size of lacewings.

Certain biological control agents feed on more than one form of prey and are called 'polyphagous'. Parasitic nematode strains consume not only slugs but also sciarid fly larvae and will attack a range of insects in the right conditions. Similarly, strains of *Verticillium lecanii* infect both whiteflies, thrips and aphids. Sometimes both chemicals and biological agents may work together to defeat a range of insects. This is referred to as integrated pest management. In this exercise you are asked to select a biological weapon for a named horticultural pest.

Aim

To apply horticultural knowledge in matching biological control predators to pests.

Apparatus

lists of natural enemy predators and prey given below
knowledge from previous exercises

Method

1 Match the following list of natural enemies to the pest problems listed.

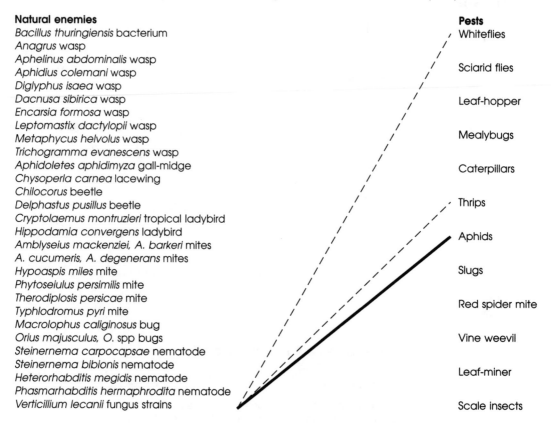

Natural enemies
Bacillus thuringiensis bacterium
Anagrus wasp
Aphelinus abdominalis wasp
Aphidius colemani wasp
Diglyphus isaea wasp
Dacnusa sibirica wasp
Encarsia formosa wasp
Leptomastix dactylopii wasp
Metaphycus helvolus wasp
Trichogramma evanescens wasp
Aphidoletes aphidimyza gall-midge
Chysoperla carnea lacewing
Chilocorus beetle
Delphastus pusillus beetle
Cryptolaemus montruzieri tropical ladybird
Hippodamia convergens ladybird
Amblyseius mackenziei, A. barkeri mites
A. cucumeris, A. degenerans mites
Hypoaspis miles mite
Phytoseiulus persimilis mite
Therodiplosis persicae mite
Typhlodromus pyri mite
Macrolophus caliginosus bug
Orius majusculus, O. spp bugs
Steinernema carpocapsae nematode
Steinernema bibionis nematode
Heterorhabditis megidis nematode
Phasmarhabditis hermaphrodita nematode
Verticillium lecanii fungus strains

Pests
Whiteflies
Sciarid flies
Leaf-hopper
Mealybugs
Caterpillars
Thrips
Aphids
Slugs
Red spider mite
Vine weevil
Leaf-miner
Scale insects

2 Use a thick line for main prey, and a dotted line for secondary prey, for example, *Verticillium lecanii* has aphids as a main prey, and thrips and white flies as a secondary prey.

Results

Indicate which pests are the main and secondary prey for each natural enemy.

Conclusions

1 Explain the term 'predator' as used in biological control.
2 Describe the integration of chemical and other control methods, when used with biological control agents.
3 Explain why a knowledge of organisms' life cycles is important in the correct application of the principles of biological control.
4 Explain what is meant by the term 'polyphagous' predator.

Useful suppliers

Most materials suggested in this publication are easily ordered locally through garden centres and chemists. The following companies have been found to be reasonable suppliers of some of the more difficult material to find.

Bunting Biological
  Control Ltd
Westwood Park
Little Horkesley
Colchester
Essex CO6 4BS
Tel: (01206) 271300
Fax: (01206) 272001

Biological control products.

Defenders Ltd
Occupation Road
Wye
Ashford
Kent TN25 5EN
Tel: (01233) 813121
Fax: (01233) 813633

Mail order biological control products.

Griffen Education
Griffen & George
Bishop Meadow Road
Loughborough
Leicestershire LE11 0RG
Tel: (01509) 233344
Fax: (01509) 231893

Educational supplier, microscopic slides and general laboratory apparatus.

Philip Harris Education
Lynn Lane
Shenstone
Lichfield
Staffordshire WS14 0EE
Tel: (01543) 480077
Fax: (01543) 480068

Educational supplier, microscopic slides.

Koppert UK Ltd
1 Wadhurst Business Park
Fairchrouch Lane
Wadhurst
East Sussex TN5 6PT
Tel: (01892) 784411
Fax: (01892) 782469

Biological control products.

Merckoquant (BDH)
Merck Ltd
Hunter Boulevard
Magna Park
Lutterworth
Leicester LE17 4XN
Tel: Freephone 0800 223344
Fax: (01455) 558589

Educational supplier, laboratory apparatus and chemicals, tetrazolium salt, pH kits.

Meteorological Office
London Road
Bracknell
Berkshire RG12 2S
Tel: (01344) 856829
Fax: (01344) 849801

Information on local weather recording stations.

Soil Survey and Land
   Research Centre
School of Agriculture, Food
   and Environment
Cranfield University
Silsoe
Beds MK45 4DT
Tel: (01525) 863000
Fax: (01525) 86325

Soil maps.

Bibliography

Adams, C.R., Bamford, K.M., and Early, M.P. (1995). *Principles of Horticulture* (2nd edition). Heinnemann.

Bagust, H. (1992). *The Gardener's Dictionary of Horticultural Terms*. Cassell Publishers Ltd.

Ball, R., and Archer, G. (1991) *Integrated Science Assignments: Communication and IT Skills*. Cambridge University Press.

Buczacki, S., and Harris, K. (1989). *Collins Guide to the Pests, Diseases and Disorders of Garden Plants*. Collins.

Clegg, C.J., and Mackean, D.G. (1994). *Advanced Biology. Principles and Applications*. John Murray Publishers Ltd.

Day, D. (Ed.). (1991). *Grower Digest 11: Biological Control – Protected Crops*. Grower Publications Ltd.

Finagin, J., and Ingram, N. (1988). *Biology for Life*. Thomas Nelson & Sons Ltd.

Fitter, R. S. R. (1985) *Wild Flowers of Britain and Northern Europe*. Collins.

Food and Agricultural Organization of the United Nations (1984). *Fertilizer and plant nutrition guide*. FAO.

Freeland, P. W. (1988). *Investigations for GCSE Biology*. Hodder & Stoughton.

Freeland, P. W. (1988). *Investigations for GCSE Biology – Teachers' Book*. Hodder & Stoughton.

Freeland, P. W. (1991). *Habitats and the Environment: Investigations*. Hodder & Stoughton.

Gratwick, M. (Ed.) (1992) *Crop Pests in the UK: Collected Edition of MAFF leaflets*. Chapman & Hall.

Hessayon, D. G. (1991). *The Lawn Expert*. PBI Publications.

Hessayon, D. G. (1992). *The Tree and Shrub Expert*. PBI Publications.

Hodgson, J. M. (1985). *Soil Survey Field Handbook* (2nd edition). Soil Survey of England and Wales (technical monograph No. 5). Bartholomew Press.

Johnson, A. T., and Smith, H. A. (1986). *Plant Names Simplified: Their Pronounciation, Derivation and Meaning*. Hamlyn Publishing Group Ltd.

Koppert B.V. (1993). *Koppert bio-journal*. Koppert B.V. The Netherlands.

Koppert B.V. (1994). *Koppert Products – With Directions for Use*. Koppert B.V. The Netherlands.

Mackean, D. G. (1983). *Experimental Work in Biology*. John Murray Publishers Ltd.

Mackean, D. G. (1989). *GCSE Biology*. John Murray Publishers Ltd.

Ministry of Agriculture, Fisheries and Food (1983). *Lime in Horticulture* (ADAS leaflet 518). HMSO.

Phillips, R. (1986). *The Photographic Guide to Identify Garden and Field Weeds*. Elm Tree.

Phillips, W. D., and Chilton, T. J. (1994). *A-Level Biology* (revised edition). Oxford University Press.

Press, B. (1993). *Bob Press's Field Guide to the Wild Flowers of Britain and Europe*. New Holland.

Roberts, M. B. V. (1980). *Biology: A Functional Approach – Students' Manual*. Thomas Nelson & Sons Ltd.

Rose, F. (1981). *The Wild Flower Key to Britain and N.W. Europe*. Warne.

Rouan, C., and Rouan, B. (1987). *Basic Biology Questions for GCSE*. The Bath Press.

Rowland, M. (1992). *Biology: University of Bath Science 16–19*. Thomas Nelson & Sons Ltd.

Sampson, C. (1995). Whitefly predator debut. *Grower* (23 February).

Scott, M. (1984). *Efford Sandbeds* (ADAS leaflet No. 847). MAFF, HMSO.

Simpkins, J., and Williams, J. I. (1989). *Advanced Biology* (3rd edition). Unwin Hyman Ltd.

Soper, R. (Ed.). (1990). *Biological Science 1 and 2*. Cambridge University Press.

Whitehead, R. (Ed.). (1995). *The UK Pesticide Guide: British Crop Protection Council, CAB International*. Cambridge University Press.

Woodward, I. (1989). Plants' Water and Climate (Inside Science No. 18). *New Scientist* (18 February).

Wye College (1995). *Wye College Biological Control Handbook*. Wye College Press.

Index